Nelson Maths

W0051251

2

Teacher's Book

Karen Morrison
Lisa Greenstein

OXFORD
UNIVERSITY PRESS

Great Clarendon Street, Oxford, OX2 6DP, United Kingdom

Oxford University Press is a department of the University of Oxford.

It furthers the University's objective of excellence in research, scholarship, and education by publishing worldwide. Oxford is a registered trade mark of Oxford University Press in the UK and in certain other countries.

© Cloud Publishing Services CC and Lisa Greenstein 2022
The moral rights of the authors have been asserted.

First published 2022

British Library Cataloguing in Publication Data

Data available

ISBN: 978-1-382-01012-2

3 5 7 9 10 8 6 4 2

Paper used in the production of this book is a natural, recyclable product made from wood grown in sustainable forests. The manufacturing process conforms to the environmental regulations of the country of origin.

Printed in Great Britain by Ashford Colour Press Ltd.

Acknowledgements

The publisher and authors would like to thank the following for permission to use photographs and other copyright material:

Cover: Matthieu Nivesse. **Photos: p9:** ann_isme/Shutterstock; **p21:** Oxford University Press.

Artwork by Lilana Perez, John Haslam, Q2A Media, Integra Software Services, Pantek Media, and Oxford University Press.

Every effort has been made to contact copyright holders of material reproduced in this book. Any omissions will be rectified in subsequent printings if notice is given to the publisher.

Cover activities

The following activities are based on the Level 2 Pupil Book and Workbook cover image. You can use these stimulus questions according to children's learning to date.

Number

Number facts: Ask the children to tell you any number facts they can see. Ask questions such as: *Are there an even or odd number of animals, trees, clouds, buildings, wheels, tractors, windows?*

Compare using <, > or =: Ask the children to write sentences using <, > or = to compare:

- the amount of sand in the red and blue truck
- the number of each animal, e.g. number of sheep = number of goats
- the number of round trees and the number of pointy trees
- the number of clouds and the number of buildings.

Can they think of a comparison question to ask their partner about the cover?

Geometry

Identify 3D and 2D shapes: Ask the children how many different 2D and 3D shapes they can see. Can they identify any 2D faces on the 3D shapes? Encourage children to describe the properties of the shapes – ask how many faces, edges and vertices they have.

Parallel lines: Ask the children if they can see any parallel lines. Are they straight or curved? How do they know they are parallel?

Line symmetry: Ask the children if there are any lines of symmetry, or shapes/objects with lines of symmetry? Give children small mirrors for support.

Contents

How Nelson Maths Works

LEVEL	PUPIL BOOK	WORKBOOKS	TEACHING SUPPORT	DIGITAL CONTENT ON OXFORD OWL
STARTER LEVEL				For all levels: • Digital versions of the Pupil Books, Workbooks and Teacher's Books • Assessment support • Parent notes • Curriculum mapping and planning guides • Vocabulary support
1				
2				
3				
4				
5				
6				

How to use Nelson Maths

Nelson Maths is a comprehensive maths programme for children aged 4–11. The course covers the following five strands: Number, Measure, Geometry, Statistics and Algebra to ensure full coverage of your chosen curriculum. The course is also full of opportunities to develop children's problem-solving skills through carefully designed questions and activities.

Nelson Maths is made up of seven levels (one for each year group) and Level 2 contains 20 units, which should ideally be taught in order. The units have been arranged in a careful progression, designed to support children to build their skills and knowledge and make connections across the different strands. Your children may progress through some units more quickly than others, but it is recommended that you spend approximately two weeks on each unit.

Children each have a **Pupil Book** and write-in **Workbook**. The main lesson activities are included in the **Pupil Book**, and the **Workbook** includes extra practice and consolidation activities. The **Teacher's Book** contains detailed teaching support for each unit, which will help you to introduce new concepts, revise prior learning, deepen children's understanding of a topic and encourage mathematical conversations and discussion.

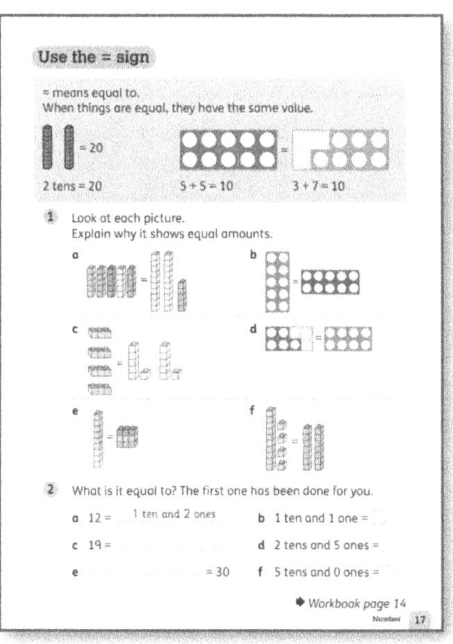
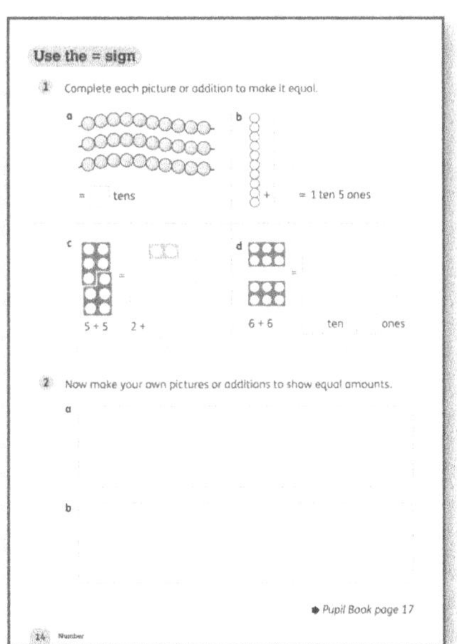

Each unit in the **Pupil Book** is supported by a corresponding unit in the **Workbook**.

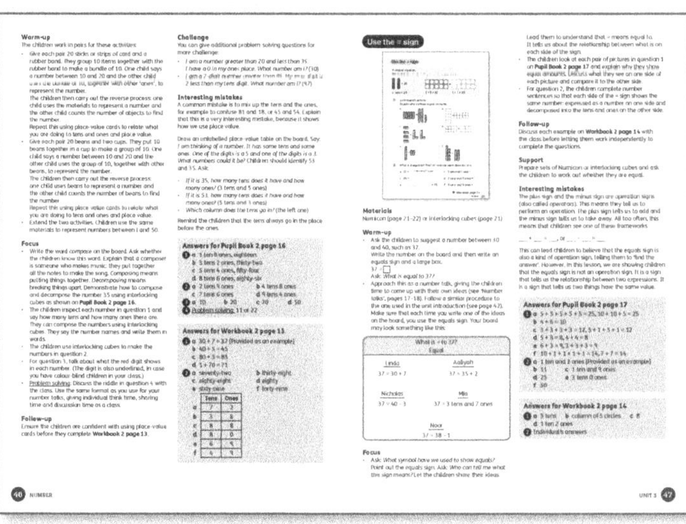

The **Teacher's Book** provides detailed teaching guidance for each unit and answers to the questions.

Read the relevant unit of the Teacher's Book before you begin teaching it. It's important to familiarise yourself with the learning objectives, key vocabulary and resources you need before each lesson. The following pages will give you a good understanding of how to use the Pupil Book, Workbook and Teacher's Book together.

In addition to the printed materials, you can also find lots of supplementary digital resources online on Oxford Owl. Please see page 9 for more information.

Using the Pupil Books

The **Pupil Book** units are designed to be worked through in order, following the discussion prompts and activities suggested in the **Teacher's Book**, and using the linked **Workbook** pages to help children to consolidate their understanding through independent practice. The following features of the **Pupil Books** are designed to help you get the most out of your lessons.

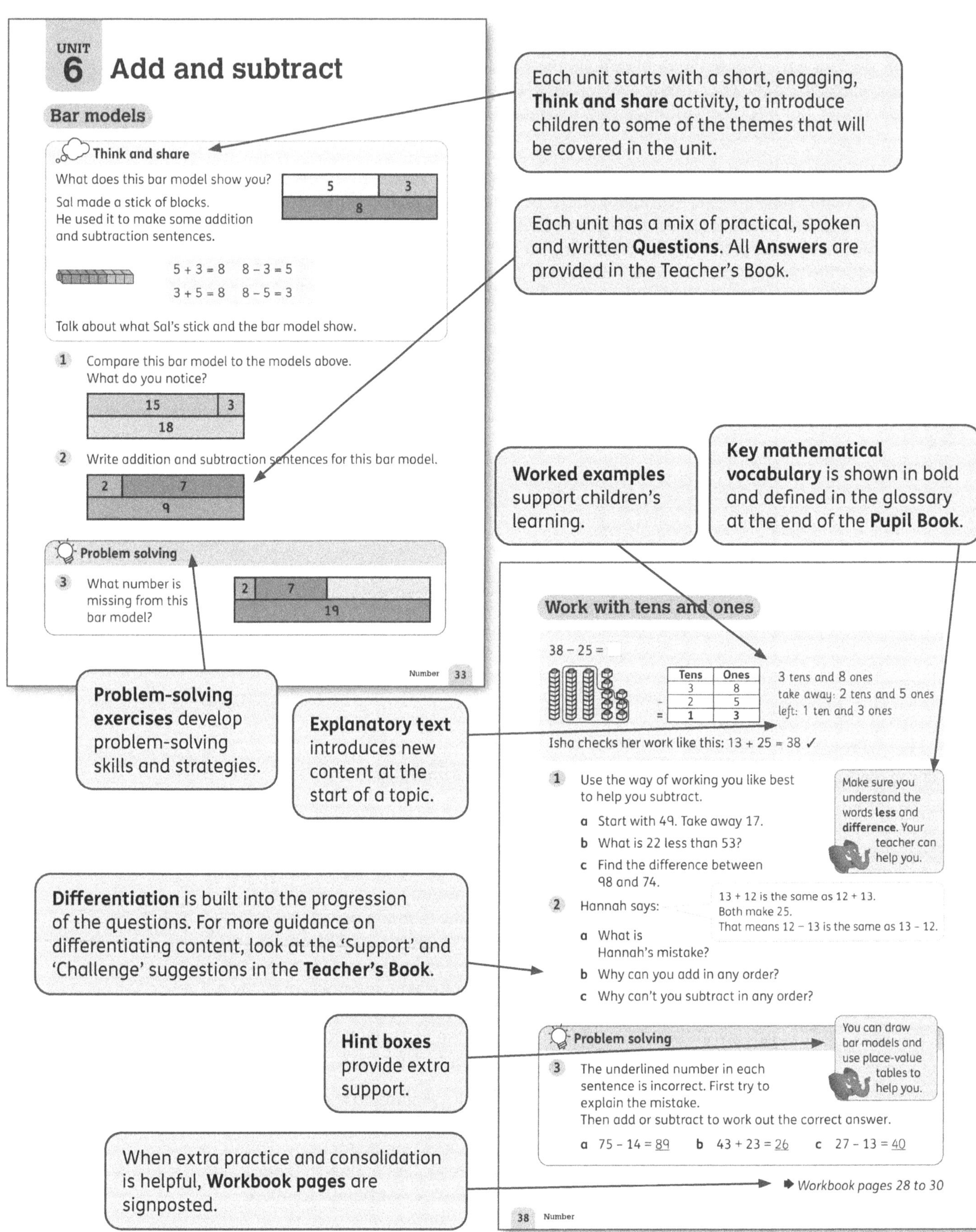

Each unit starts with a short, engaging, **Think and share** activity, to introduce children to some of the themes that will be covered in the unit.

Each unit has a mix of practical, spoken and written **Questions**. All **Answers** are provided in the Teacher's Book.

Key mathematical vocabulary is shown in bold and defined in the glossary at the end of the **Pupil Book**.

Worked examples support children's learning.

Problem-solving exercises develop problem-solving skills and strategies.

Explanatory text introduces new content at the start of a topic.

Differentiation is built into the progression of the questions. For more guidance on differentiating content, look at the 'Support' and 'Challenge' suggestions in the **Teacher's Book**.

Hint boxes provide extra support.

When extra practice and consolidation is helpful, **Workbook pages** are signposted.

UNIT 1 — Think maths

How I feel about maths

Think and share

Read what these pupils said:

How do you think each pupil feels about making a mistake?

1. When you do maths, you use your brain. There is no such thing as a maths brain. Everyone's brain looks a bit like this:

A human brain

   a. Which shape is most like the shape of the br

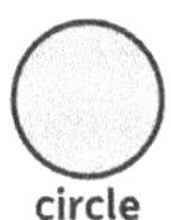  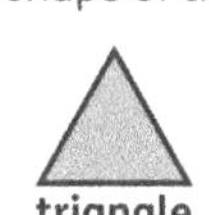

   circle   square   triangle

   b. What kind of lines does it have?

Mixed practice 1

1. Teia showed the number 6 like this, using counters. Draw two different = ways you can arrange counters to show the number 6.

2. The first tin contains beans.

   a. Which tins are second, third and fourth in the row?

   b. In which positions are the coconut milk and the corn?

   c. Write two more sentences about the positions of the other tins.

3. Write the number names for these numbers. Then write how many tens and how many ones.

   a. 48   b. 92   c. 80   d. 39   e. 25

4. Write these numbers in order from smallest to greatest.

   a. 11, 18, 4, 23   b. 53, 32, 23, 35

5. What 2D shape am I? Draw the shape and write what it is called.

   a. I am a round shape with no corners.

   b. I have four square corners, but my sides are not all equal.

6. Write the name of each 3D shape.

   a.  b. 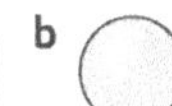 c.  d.  e.  f. 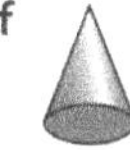

7. Draw or write your own repeating pattern using three different shapes. Explain how your pattern repeats.

Using the Workbooks

The **Workbooks** provide essential practice to help children consolidate their learning. They provide new questions and activities linked to the concepts introduced in the **Pupil Books**. They are designed for independent use, making them ideal for homework or additional classwork. We recommend that children complete each page in the **Workbook** after they have completed the corresponding lesson in the **Pupil Book**. When extra practice and consolidation is helpful, **Workbook** pages are signposted at the end of the Pupil Book lesson.

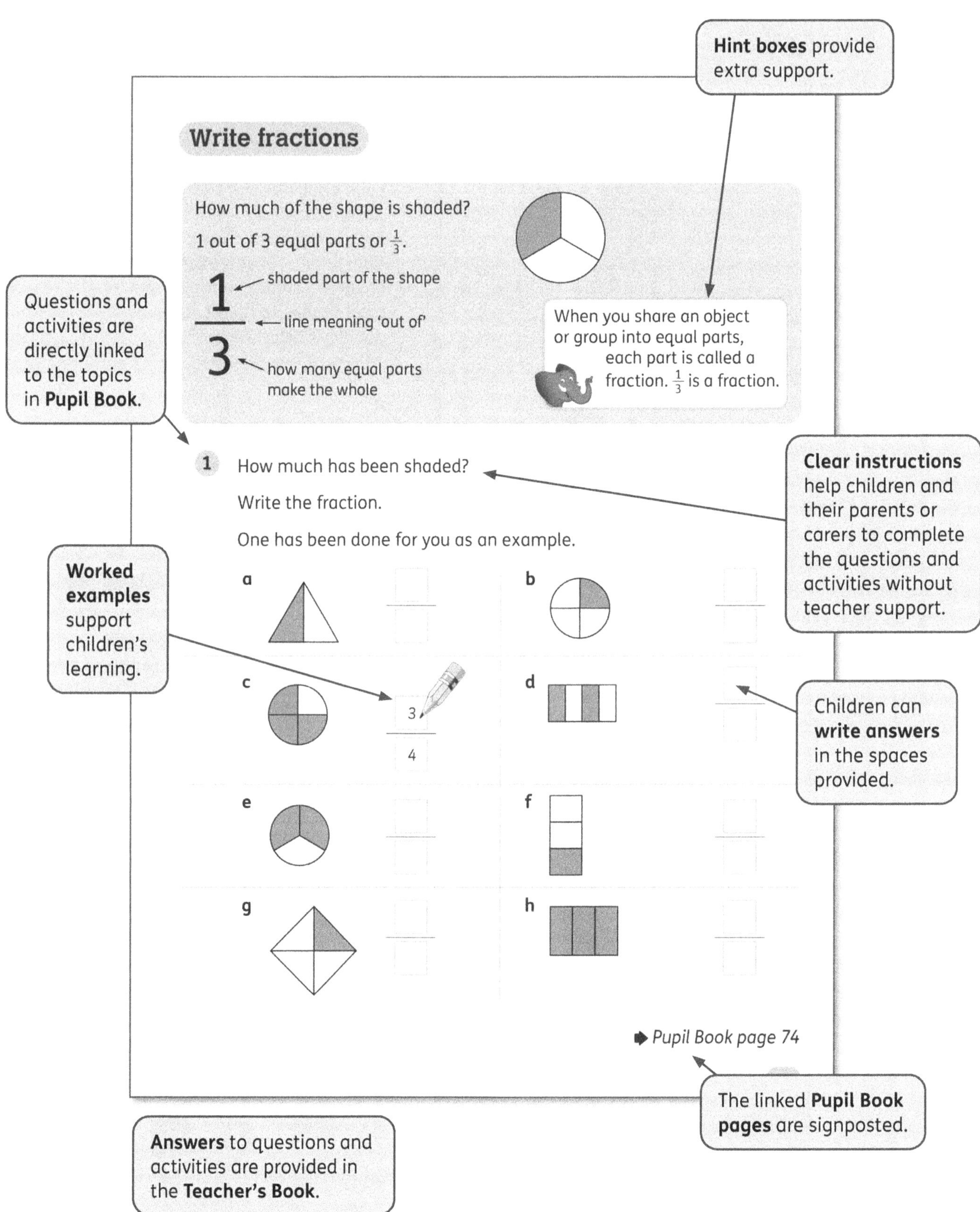

Using the Digital Content

This edition of *Nelson Maths* is supported by additional digital resources available online on **Oxford Owl**. These include:

- Digital versions of the Pupil Books, Workbooks and Teacher's Books to support planning and for front of class display
- A set of printable assessments with a progress tracking tool and mark scheme
- Guidance for you to share with children's parents or carers about their child's learning, including support for homework and ideas for incorporating maths into everyday life
- Vocabulary support
- Curriculum correlation charts
- Planning support

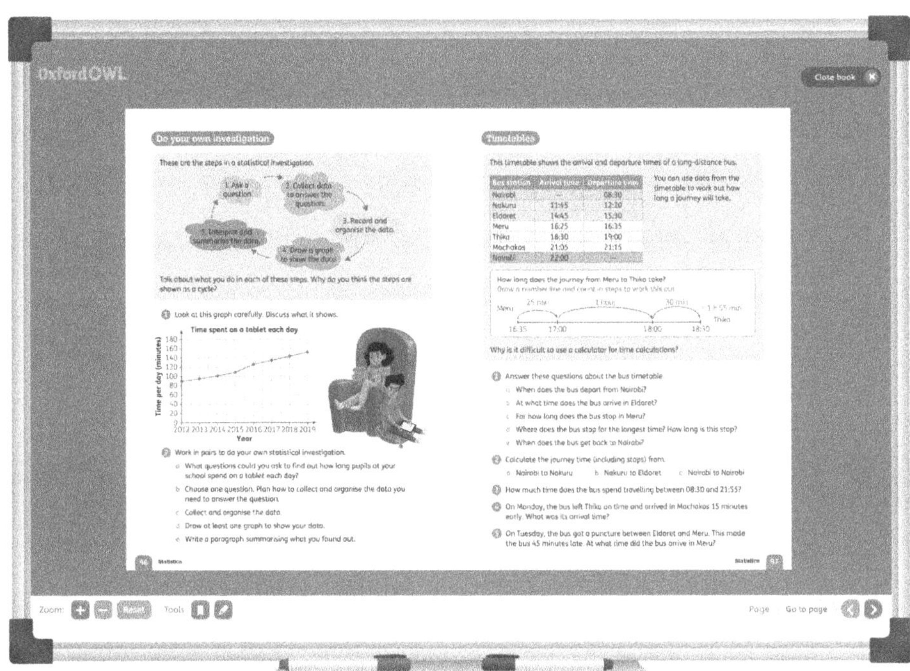

Using the Teacher's Book

The **Teacher's Book** provides step-by-step lesson notes for each unit in the **Pupil Book** and **Workbook**, including learning objectives, answers to the questions activities and suggestions for additional challenge and support.

Before you start teaching the units, take some time to familiarise yourself with these useful sections at the start of the **Teacher's Book**. These four sections are designed to help you get the most out of the *Nelson Maths* teaching materials:

1. **Introduction** – find out more about the mathematical strands, skills and fundamental principles that underpin *Nelson Maths*. Get ideas and support for teaching maths in an inclusive way that acknowledges individual differences and human diversity.

2. **Teaching approach** – guidance for teaching *Nelson Maths* effectively, including:

- how to develop robust learners with a growth mindset and an understanding of the importance of making mistakes
- the importance of the do-talk-record model that underpins the lessons
- using engaging teaching methods, from investigations and number talks, to exploring maths in real life contexts.

3. **An environment for exploration and play** – practical support on organising your classroom space; approaching whole-class, group and individual activities; and using manipulatives and games to deepen children's learning.

4. **Activity bank of warm-ups and mental maths** – a bank of engaging mental maths, support, and consolidation activities to use with your class.

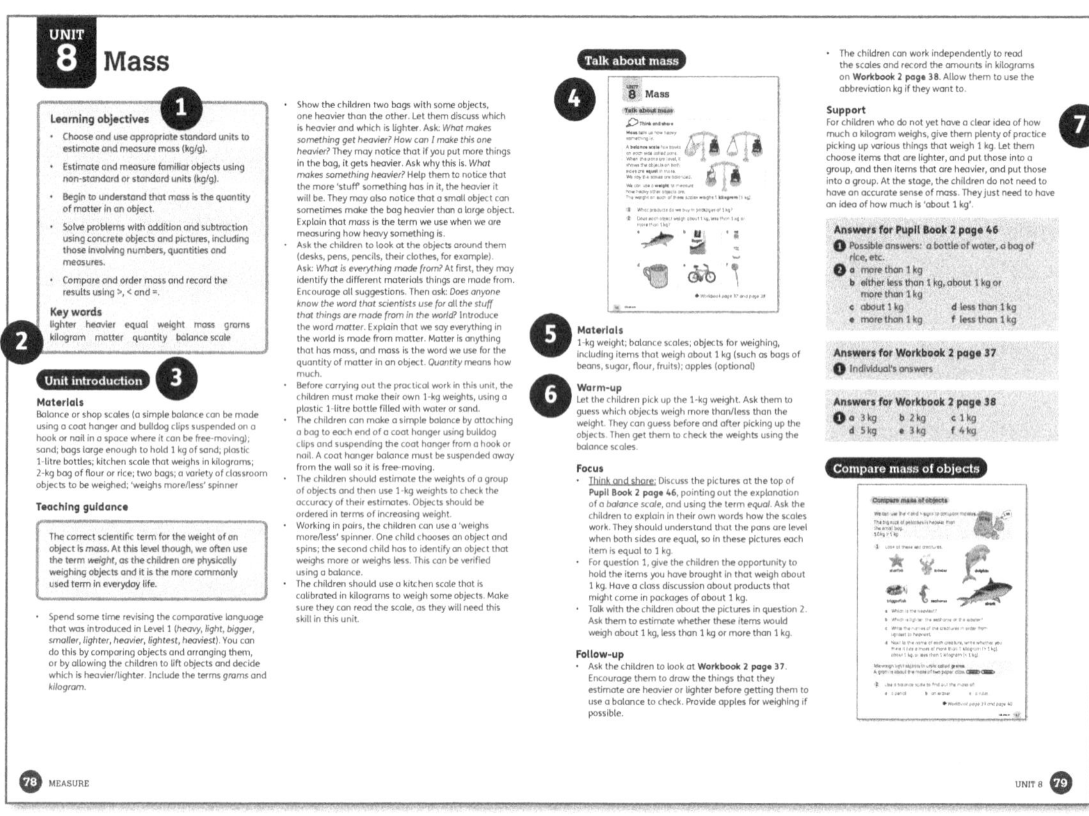

UNIT 8 — Mass

Learning objectives
- Choose and use appropriate standard units to estimate and measure mass (kg/g).
- Estimate and measure familiar objects using non-standard or standard units (kg/g).
- Begin to understand that mass is the quantity of matter in an object.
- Solve problems with addition and subtraction using concrete objects and pictures, including those involving numbers, quantities and measures.
- Compare and order mass and record the results using >, < and =.

Key words
lighter heavier equal weight mass grams kilogram matter quantity balance scale

Unit introduction

Materials
Balance or shop scales (a simple balance can be made using a coat hanger and bulldog clips suspended on a hook or nail in a space where it can be free-moving); sand; bags large enough to hold 1 kg of sand; plastic 1-litre bottles; kitchen scale that weighs in kilograms; 2-kg bag of flour or rice; two bags; a variety of classroom objects to be weighed; 'weighs more/less' spinner

Teaching guidance

The correct scientific term for the weight of an object is mass. At this level though, we often use the term weight, as the children are physically weighing objects and it is the more commonly used term in everyday life.

- Spend some time revising the comparative language that was introduced in Level 1 (heavy, light, bigger, smaller, lighter, heavier, lightest, heaviest). You can do this by comparing objects and arranging them, or by allowing the children to lift objects and decide which is heavier/lighter. Include the terms grams and kilogram.
- Show the children two bags with some objects, one heavier than the other. Let them discuss which is heavier and which is lighter. Ask: What makes something get heavier? How can I make this one heavier? They may notice that if you put more things in the bag, it gets heavier. Ask why this is. What makes something heavier? Help them to notice that the more 'stuff' something has in it, the heavier it will be. They may also notice that a small object can sometimes make the bag heavier than a large object. Explain that mass is the term we use when we are measuring how heavy something is.
- Ask the children to look at the objects around them (desks, pens, pencils, their clothes, for example). Ask: What is everything made from? At first, they may identify the different materials things are made from. Encourage all suggestions. Then ask: Does anyone know the word that scientists use for all the stuff that things are made from in the world? Introduce the word matter. Explain that we say everything in the world is made from matter. Matter is anything that has mass, and mass is the word we use for the quantity of matter in an object. Quantity means how much.
- Before carrying out the practical work in this unit, the children must make their own 1-kg weights, using a plastic 1-litre bottle filled with water or sand.
- The children can make a simple balance by attaching a bag to each end of a coat hanger using bulldog clips and suspending the coat hanger from a hook or nail. A coat hanger balance must be suspended away from the wall so it is free-moving.
- The children should estimate the weights of a group of objects and then use 1-kg weights to check the accuracy of their estimates. Objects should be ordered in terms of increasing weight.
- Working in pairs, the children can use a 'weighs more/less' spinner. One child chooses an object and spins; the second child has to identify an object that weighs more or weighs less. This can be verified using a balance.
- The children should use a kitchen scale that is calibrated in kilograms to weigh some objects. Make sure they can read the scale, as they will need this skill in this unit.

Talk about mass

Materials
1-kg weight; balance scales; objects for weighing, including items that weigh about 1 kg (such as bags of beans, sugar, flour, fruits); apples (optional)

Warm-up
Let the children pick up the 1-kg weight. Ask them to guess which objects weigh more than/less than the weight. They can guess before and after picking up the objects. Then get them to check the weights using the balance scales.

Focus
- Think and share: Discuss the pictures at the top of Pupil Book 2 page 46, pointing out the explanation of a balance scale, and using the term equal. Ask the children to explain in their own words how the scales work. They should understand that the pans are level when both sides are equal, so in these pictures each item is equal to 1 kg.
- For question 1, give the children the opportunity to hold the items you have brought in that weigh about 1 kg. Have a class discussion about products that might come in packages of about 1 kg.
- Talk with the children about the pictures in question 2. Ask them to estimate whether these items would weigh about 1 kg, less than 1 kg or more than 1 kg.

Follow-up
- Ask the children to look at Workbook 2 page 37. Encourage them to draw the things that they estimate are heavier or lighter before getting them to use a balance to check. Provide apples for weighing if possible.
- The children can work independently to read the scales and record the amounts in kilograms on Workbook 2 page 38. Allow them to use the abbreviation kg if they want to.

Support
For children who do not yet have a clear idea of how much a kilogram weighs, give them plenty of practice picking up various things that weigh 1 kg. Let them choose items that are lighter, and put those into a group, and then items that are heavier, and put those into a group. At the stage, the children do not need to have an accurate sense of mass. They just need to have an idea of how much is 'about 1 kg'.

Answers for Pupil Book 2 page 46
1. Possible answers: a bottle of water, a bag of rice, etc.
2. a more than 1 kg
   b either less than 1 kg, about 1 kg or more than 1 kg
   c about 1 kg   d less than 1 kg
   e more than 1 kg   f less than 1 kg

Answers for Workbook 2 page 37
1. Individual's answers

Answers for Workbook 2 page 38
1. a 3 kg   b 2 kg   c 1 kg
   d 5 kg   e 3 kg   f 4 kg

Compare mass of objects

78 MEASURE · UNIT 8 79

- You may want to give the following simple example first and present it as a number talk (see Number talks, pages 17–18). This object weighs 2 kg. This object weighs 3 kg. How much more does the second object weigh than the first? How do you work it out?
- Suggestions could include counting back along a number line and subtracting to find the difference.
- Problem solving: For question 3, the children need to discuss which is heavier – the melon or the weights – and what this could tell us. We don't have any exact weights, but we can make statements using 'more than' and 'less than'.

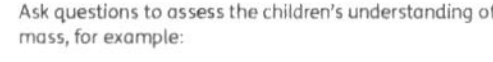

Challenge
Set some additional problems, such as these:
- Ask the children to look at the picture of the mangoes on Pupil Book 2 page 48. Say: This picture shows 12 mangoes. The tray of mangoes weighs 14 kg. About how much do you think each mango weighs? (Just over 1 kg)
- Ask the children to use the masses of the animals on Workbook 2 page 40. Give them these instructions: Draw a number line. Label the left end heaviest and the right end lightest. Mark points on the line to position the masses of the animals correctly.

Interesting mistakes
Some children may make the mistake of thinking that asking how much heavier or how much lighter determines the operation as addition or subtraction. Give more simple examples to demonstrate the concept of difference: For example, say: A melon weighs 5 kg and a pumpkin weighs 7 kg. Show this using, say 5 counters and 7 counters. Ask:
- How many more counters are on this side? (2)
- How many counters fewer are on this side? (2)
- What is the difference between 5 and 7? (2)
- How much more does the pumpkin weigh? (2 kg) How much less does the melon weigh than the pumpkin? (2 kg) Why do both questions give the same answer? (The difference between the melon and pumpkin is 2 kg. The order does not matter.)

Answers for Pupil Book 2 page 48
1. a oranges, mangoes   b grapes
   c 5 kg   d 11 kg
2. a 15 kg   b 24 kg   c 19 kg
Problem solving:
3. a They have the same mass.
   b Each weight is 1 kg.
   c Each weight is ½ kg.
   d It weighs 2 kg.

End-of-unit check
Ask questions to assess the children's understanding of mass, for example:
- What does kg stand for? (kilogram)
- Can you arrange these weights in increasing order of mass? (Give the children five different weights.)
- How do you know which object is heavier on a balance scale? (The heaviest item will make that side of the scale lower.)
- A box of bricks has a mass of 12 kg. A box of tiles has a mass of 9 kg. Which is heavier? (The box of bricks) How much heavier? (3 kg)
- A small packet has a mass of 4 kg. A larger packet has a mass of 7 kg. What is their mass altogether? (11 kg)
- Jossie weighs 10 kg more than Mika. If Mika weighs 23 kg, how much does Jossie weigh? (33 kg)

Mixed practice 1

Answers to Mixed practice 1 on Pupil Book 2 pages 49–50
1. Possible answers: a row of 6 counters; a row of 5 counters + 1 counter; a row of 4 counters + a row of 2 counters
2. a second: tomatoes   third: pears   fourth: pineapples
   b coconut milk: fifth   corn: ninth
   c Individual's answers
3. a forty-eight; 4 tens 8 ones
   b ninety-two; 9 tens 2 ones
   c eighty; 8 tens 0 ones
   d thirty-nine; 3 tens 9 ones
   e twenty-five; 2 tens 5 ones
4. a 4, 11, 18, 23   b 23, 32, 35, 53
5. a circle   b rectangle
6. a cube   b sphere   c cuboid   d pyramid   e cylinder   f cone
7. Individual's answers. The explanation should use the letters A, B, C, for example: ABC, ABBC (to match the sequence of the three shapes).
8. 5, 10, 15, 20, 25, 30, 35, 40, 45, 50
9. a 5 cm > 3 cm   b 2 cm < 6 cm   c 1 cm < 2 cm
10. 25 cm
11. 25 cm
12. a The height of Kevin's tower   b 12 cm
13. a less than 1 kg   b equal to 1 kg   c more than 1 kg

UNIT 8 81

---

1 **Learning objectives** are listed at the beginning of each unit.

2 **Key words** match the important words highlighted in bold in the **Pupil Book**.

3 **Unit introduction** activities introduce the topics in the unit, engage children and find out what they already know.

4 The **linked Pupil Book and Workbook pages** are signposted.

5 **Materials** children need for the lesson are listed.

6 Most lessons are broken down into three stages:
- **Warm-up** – introduce the topic and unlock prior learning
- **Focus** – develop a firm understanding of concepts and skills by working through the **Pupil Book**
- **Follow-up** – recap the key learning points and extend the learning with further activity suggestions in the **Workbook**.

7 Where relevant, **Challenge** and **Support** sections give guidance on deepening the learning or making it more accessible, so that every child is catered for.

8 It is important for children to see mistakes as part of their learning journey. The **Interesting mistakes** section highlights common errors and misconceptions that children may have, along with what to look for and advice on how to support children.

9 **Answers** for **Pupil Book** and **Workbook** questions are provided to make marking easy.

10 An **End-of-unit check** allows you to assess how well children have understood the concepts in each unit, so you can monitor children's understanding throughout the course. A set of formative assessment questions or a closing activity are provided with answers.

Level 2 Scope and sequence

Unit	Unit title	Strand	Learning objectives	Key words
1	Think maths	Think maths	• Establish a classroom environment conducive to thinking and working mathematically • Set up positive norms • Establish key messages of growth-mindset mathematics	brain, synapse, mistake, question
2	Working with numbers	Number	• Recognise a number of objects presented in different ways without counting • Estimate and count objects and people to 100 • Recognise a number of objects presented in unfamiliar patterns up to 10, without counting • Recognise and use ordinal numbers (first, second, third, and so on) • Read and write numbers to at least 100 in numerals and words. • Identify, represent and estimate number using different representations, including number lines • Count in steps of 2, 3, and 5 from 0, and in tens from any number, forward and backward	count, number, tens, ones, number line, estimate, how many, arrange, order, number names (zero to one hundred), ordinal numbers
3	Place value	Number	• Recognise the place value of each digit in a 2-digit number (tens, ones) • Understand and explain the value of each digit in a 2-digit number • Use place value and number facts to solve problems • Compare and order numbers from 0 to 100 • Use <, > and = signs • Compose, decompose and regroup 2-digit numbers using tens and ones	place value, place-value table, tens, ones, digit, even, odd, compose, decompose, < sign, > sign
4	2D and 3D shapes	Geometry	• Identify and describe the properties of 2D shapes, including the number of sides • Identify, describe, sort and name 3D shapes by their properties, including the number of edges, vertices and faces • Identify 2D shapes on the surface of 3D shapes (for example, a circle on a cylinder and a triangle on a pyramid) • Compare and sort common 2D and 3D shapes and everyday objects • Understand that a circle has a centre and any point on the boundary is at the same distance from the centre	2D shape, 3D shape, sphere, cube, cuboid, pyramid, cone, prism, cylinder, edge, face, polygon/regular polygon, properties, side, vertex, vertices, circle, square, triangle, rectangle, corner, boundary, centre, length
5	Patterns and sequences	Number	• Recognise, describe and extend numerical sequences (0 to 100) • Order and arrange combinations of mathematical objects in patterns and sequences • Count in steps of 2, 3 and 5 from 0, and in tens from any number forward and backward	pattern, shape, repeat, repeating shapes, growing patterns, core, terms, rules
6	Add and subtract	Number	• Recall and use addition and subtraction facts to 20 fluently and derive and use related facts up to 100 • Solve problems with addition and subtraction: using concrete objects and pictorial representations, including those involving numbers, quantities and measures; applying their increasing knowledge of mental and written methods • Add and subtract using concrete objects, pictorial representations and mentally: 2-digit numbers and ones; 2-digit numbers and tens; two 2-digit numbers; adding three 1-digit numbers • Show that addition of numbers can be done in any order (commutative), but that subtraction cannot • Recognise and use the inverse relationship between addition and subtraction and use this to check calculations and solve missing number problems • Count in steps of 2, 3, and 5 from 0, and in tens from any number, forward and backward • Recognise the place value of each digit in a 2-digit number (tens, ones) • • Identify, represent and estimate numbers using different representations, including the number line	add, subtract, total, sum, order, count on, pair, round

Unit	Unit title	Strand	Learning objectives	Key words
7	Length	Measure	• Solve problems with addition and subtraction using concrete objects and pictures, including those involving numbers, quantities and measures • Estimate and measure lengths of objects with non-standard or standard units • Choose and use appropriate standard units to estimate and measure length/height in any direction (m/cm) • Compare and order lengths and record the results using >, < and = • Understand that length is a fixed distance between two points • Draw and measure lines using standard units • Understand that the markings on a ruler are a type of scale and connect them to a number line where intermediate points have value • Read from numbered scales on a ruler, starting at 0 • Use the language of approximation	length, long, wide, high, ruler, metre stick, standard units, non-standard units, centimetre, metre, estimate, compare
8	Mass	Measure	• Choose and use appropriate standard units to estimate and measure mass (kg/g) • Estimate and measure familiar objects using non-standard or standard units (kg/g) • Begin to understand that mass is the quantity of matter in an object • Solve problems with addition and subtraction using concrete objects and pictures, including those involving numbers, quantities and measures • Compare and order mass and record the results using >, < and =	lighter, heavier, equal, weight, mass, grams, kilogram, matter, quantity, balance scale
9	Lists and tables	Statistics	• Interpret and construct simple tally charts • Ask and answer simple questions by counting the number of objects in each category and sorting the categories by quantity • Ask and answer questions about totalling and comparing categorical data • Ask and answer simple questions by counting the number of objects in each category	data, table, row, column, tally mark, tally table, Venn diagram, Carroll diagram, categories, even, odd
10	Show data	Statistics	• Interpret and construct simple pictograms, tally charts, block diagrams and simple tables • Ask and answer simple questions by counting the number of objects in each category and sorting the categories by quantity • • Ask and answer questions about totalling and comparing categorical data	chart, pictogram, block diagram, key, row column
11	Multiply	Number	• Recall and use multiplication facts for the 2, 5 and 10 multiplication tables, including recognising odd and even numbers • Calculate mathematical statements for multiplication within the multiplication tables and write them using the multiplication ($\times$) and equals (=) signs • Solve problems involving multiplication, using materials, arrays, repeated addition, mental methods and multiplication facts, including problems in contexts • Show that multiplication of two numbers can be done in any order (commutative)	multiply, many times, groups, equal, repeated addition, times, skip count, product, multiplication sign, array, times tables
12	Divide	Number	• Recall and use multiplication and division facts for the 2, 5 and 10 multiplication tables, including recognising odd and even numbers • Calculate mathematical statements for multiplication and division within the multiplication tables and write them using the multiplication ($\times$), division ($\div$) and equals (=) signs • Solve problems involving multiplication and division, using materials, arrays, repeated addition, mental methods, and multiplication and division facts, including problems in contexts • Show that multiplication of two numbers can be done in any order (commutative) and division of one number by another cannot • Understand division as: sharing (number of items per group); grouping (number of groups); repeated subtraction	divide, equal groups, shared equally, fact family
13	Fractions	Number	• Recognise, find, name and write fractions $\frac{1}{3}$, $\frac{1}{4}$, $\frac{2}{4}$ and $\frac{3}{4}$ of a length, shape, set of objects or quantity • Understand that fractions can act as operators (be interpreted as division) • Write simple fractions for example, $\frac{1}{2}$ of 6 = 3 and recognise the equivalence of $\frac{2}{4}$ and $\frac{1}{2}$	fraction, whole, part, equal, unequal, equivalent, half, halves, quarter, three-quarters, thirds

Unit	Unit title	Strand	Learning objectives	Key words
14	Time	Measure	• Know the number of minutes in an hour and the number of hours in a day • Compare and sequence intervals of time • Interpret and use simple tables (in the form of calendars)	day, week, month, year, leap year, names of months, names of days, first, second, third, fourth (up to twelfth), calendar, dates
15	Possible outcomes	Statistics	• Order and arrange combinations of mathematical objects in patterns and sequences • Conduct probability experiments with two outcomes, and present and describe the results	pattern, possible, impossible, likely, certain, sure, maybe, outcome, experiment, results
16	Symmetry	Geometry	• Identify a line of symmetry on 2D shapes and patterns • Sketch the reflection of a 2D shape in a vertical mirror line including where the mirror line is the edge of the shape	triangle, square, rectangle, pentagon, hexagon, sides, angles, reflection, line symmetry, symmetrical, mirror line, line of symmetry
17	Capacity and temperature	Measure	• Understand that capacity is a measure of how much liquid a container can contain or hold • Choose and use appropriate standard units to estimate and measure temperature (°C) and capacity (litres/ml) using thermometers and measuring vessels • Solve problems with addition and subtraction using quantities and measures • Compare and order volume/capacity and record the results using >, < and =	capacity, container, more, less, litre (ℓ), millilitre (ml), pour, temperature, hot, cold, warm, cool, degrees Celsius (°C), thermometer, measure
18	More about time	Measure	• Read and record time to five minutes in digital notation (12-hour) and on analogue clocks • Tell and write the time to five minutes, including quarter past/to the hour and draw the hands on a clock face to show these times • Know the number of minutes in an hour and the number of hours in a day	clock, minute, hour, quarter hour, quarter to/past, digital
19	Position and movement	Geometry	• Use mathematical vocabulary to describe position, direction and movement, including movement in a straight line and distinguishing between rotation as a turn and in terms of right angles for quarter, half and three-quarter turns (clockwise and anti-clockwise)	right, left, up, down, straight, direction, turn, quarter turn, half turn, three-quarter turn, full turn, right angle, forwards, backwards, straight, clockwise, anti-clockwise
20	Money	Measure	• Recognise value and money notation used in local currency • Recognise and use symbols for pounds (£) and pence (p) • Combine amounts to make a particular value • Find different combinations of coins that equal the same amounts of money • Solve simple problems in a practical context involving addition and subtraction of money of the same unit, including giving change	coins, notes, money, currency, cash, value, combination, worth, change

Section 1 Introduction

This Teacher's Book is designed to support the component parts of *Nelson Maths* Level 2.

Sections 1-4 of this Teacher's Book (pages 14 to 31) provide background and reference information, tips on classroom organisation, and a bank of activities and games to use for teaching practically. Section 5 provides lesson notes for the Pupil Book and Workbook unit by unit and page by page.

Strands and skills

In Levels 1 to 6, the content is divided into strands:
- Number – numbers and the number system, and calculating
- Measure – money, length, mass and capacity, and time
- Geometry – shape, position and movement
- Statistics – organising, categorising and representing data
- Algebra – use of formulae, variables and equations

In this course, problem solving is not treated as a separate strand but is integrated into the materials at all levels.

Fundamental principles

This series makes the following assumptions about the teaching of mathematics:
- Children need concrete (practical, hands-on) experiences in order to acquire sound mathematical understanding. Like adults, children learn best when they investigate and make discoveries for themselves.
- Children refine their understanding and develop conceptual structures by talking about their own thinking and what they have done.
- Individual children develop at different rates. While some will find certain elements of mathematics difficult, others will understand them quickly.
- Children learn in a variety of ways, so mathematics teaching should provide a rich and wide variety of experiences. Children need plenty of opportunities to apply what they have learnt and to relate their mathematics work to other areas of the curriculum and their daily lives.
- Children will become more mathematically able if they are allowed to develop reliable personal ways of working. The formal recording used by mathematicians comes with time and experience, and we cannot expect young children to understand it or use it themselves before they are ready to do so. The conventions (formal methods) should be taught only once children are confident in their own knowledge, concepts and skills.

- Children learn mathematics most effectively when they enjoy what they are doing and when they can see the relevance of what they are learning.

This course reflects current thinking about the most effective ways of teaching and learning mathematics at primary level. It recognises the professionalism of the teacher, and acknowledges that teachers are the best judges of which experience(s) are most appropriate at different stages for the children in their classes. For this reason, this course does not impose a rigid, inflexible structure. Instead, it provides a wide variety of practical activities and games linked to clearly defined purposes and objectives. The teacher selects according to the needs of their classes, groups and individuals.

Individual differences and inclusivity

Everyone learns at their own pace, and in different ways. We also need to recognise that each child enters the classroom with their own experience, their own identity, and their own hopes, fears and curiosities.

This course recognises individual differences and aims to give children the chance to explore the world of mathematics and solve problems in their own way. To achieve this, we believe that each child needs to feel that maths is for them too. Therefore, examples and illustrations need to include a wide range of children of different genders and ethnic, cultural and linguistic backgrounds and should not promote stereotypes.

- When giving children additional maths problems to solve:
  - Use examples of girls doing things that are traditionally considered 'for boys'.
  - Use examples of dads as primary child carers as well as working mothers and female leaders.

- Be aware of children's differences and acknowledge different bodies: not all children have ten fingers and ten toes, and some may not be able to walk or run as easily as others.
- Avoid comparisons of height or weight that some children may experience as body shaming.
- Celebrate slower workers in your classroom as much as faster workers. Praise reasoning and focused working rather than speed. Never focus on speed testing or 'who can find the answer the fastest'. Inclusivity means acknowledging that our slower, deeper thinkers are some of our best mathematical minds too. Maths is not about speed. It is about reasoning.

- Praise different ways of working. Some children need to move and fidget and play in order to solve problems. Include those who think visually or rely on practical manipulatives and drawings to solve their problems.
- Struggle and confusion are key parts of the process of maths learning. Never dismiss a child's struggle as evidence that they cannot do maths or that they are not mathematically minded.
- When you are planning classroom activities, bear in mind the abilities and needs of your group and change or adapt activities as necessary. For example, it may be more appropriate to count sets of ten counters, or use ten frames (see page 21) than to use fingers for counting. When activities expect children to identify, sort or match colours, adapt these activities for children who have colour-blindness, by focusing on pattern rather than colour.

Section 2 Teaching approach

A growth mindset

In recent years, scientists have made important discoveries about how the brain learns and grows. A key discovery is the notion of neuroplasticity or brain plasticity. Brain plasticity is the brain's ability to change connections and neural pathways. Put simply, this means that the brain can learn and grow throughout a person's life.

These discoveries have led many educators to talk about 'growth mindset' versus 'fixed mindset'. People with a fixed mindset believe their abilities are determined and fixed, for example by genetics or by factors outside their control. A fixed mindset can lead to fears of making mistakes. Children with fixed mindsets may believe that they are either 'good at' or 'bad at' maths, and may easily get frustrated when they find problems challenging. We need to help children to understand that our maths ability changes all the time as we learn and grow. By getting children to understand and believe that growth, development and improvement are always possible, we can foster a 'growth mindset', in which children are not afraid of tackling difficult problems.

Getting rid of maths myths

Every child has the potential to learn maths, irrespective of their existing level of mathematical ability. The ability to learn maths is also not related to other traits such as gender, race or ethnicity. There is no such thing as a 'maths person'. It is extremely important that teachers understand that all children are able to learn to think mathematically. Maths is not a special subject and does not require any special natural 'talent'. Some children may grasp concepts more quickly than others – just like in other subjects – but giving them the time to develop their understanding allows all children to achieve highly in mathematics.

The importance of mistakes

Scientists have also studied the brains of people attempting to solve problems. They found something that might seem surprising – that more neural connections and pathways form when we make mistakes than when we get things right! Although getting correct answers is usually praised by teachers, it is key that we also start understanding the importance of mistakes. Mistakes offer a valuable opportunity to explore why and how a concept makes sense. You can explore mistakes in a productive way with your class by:

- frequently reminding the children that making mistakes shows we are learning something new
- ensuring that the children know that mistakes (and questions) are welcome in your classroom
- discussing mistakes with interest and curiosity
- instead of always asking a question that invites a correct answer, sometimes phrasing questions to ask the children to identify the *incorrect* answers and then explain why they don't work.

You can read more about growth mindset in books by Carol Dweck and Jo Boaler, or search for these online.

The do–talk–record model

The learning framework for this course can be summarised as: *do–talk–record*. In other words, the starting point is always to explore and investigate concepts. Talking, discussing and explaining follows. Finally, written work can record and represent the concepts that the child has already explored practically and in discussion.

Doing

The first step in developing and understanding concepts is to let children:

- handle and manipulate apparatus
- play games
- investigate patterns and rules using real objects
- model situations and problems using real objects.

Children need time to play and explore in this way before they are expected to communicate about their work.

Talking

Children can make sense of what they have been doing by discussing:

- what they have done
- why they have done it
- what they have found out.

This allows them to generalise concepts and ideas from their particular experiences. The teacher's role is to create situations for discussion and to ask open-ended but directed questions. Most of the activities in this Teacher's Book will help you to facilitate discussion and will encourage the children to listen to each other and experiment with different ways of thinking about and solving problems.

Recording

In Level 2, children are not likely to have refined the skills and knowledge or developed the use of strategies for solving problems. They will need to use informal and very personal methods (jottings) of recording steps in a process, or keeping track of what they have done.

Jottings are an important step in moving towards non-standard methods of calculation (such as diagrams and jumps on a number line). These will give the children a foundation for more concise standard written methods of recording. It is very important that you allow, and in fact encourage, the children to make use of jottings as they work. Here are some possible ways of doing this in the classroom:

- Do jottings of your own as you work out solutions. For example, if you are demonstrating how to add 7 + 8 + 3 you might jot the following on the board to show your thinking:
  7 + 3 → 10
  10 + 8 → 18
- Talk through the jottings as you make them. Say, for example: *It's easier to add tens, so I'll add 7 and 3 first.* This modelling process helps the children to see that jottings are important and useful.
- Make space for jottings in the children's exercise books. You can reinforce the importance of jottings as a means of showing your working by encouraging the children to jot as they work. If you only allow jotting on scrap paper, the children may think it is not as important or valuable as the 'real' work in their book.
- Do activities where jotting is the point of the activity. For example, ask the children to represent a calculation visually in as many ways as possible, or ask them to work out problems where they will need to jot down steps to keep track of the process (such as *How many different ways can you pay for an item costing … using coins only?*).
- Ask the children to share their jottings and compare them to show that there are different methods of working. This can help the children to see that some strategies are more efficient than others, which can refine their own thinking. For example, in the coin task above you may find that some children draw coin combinations, others list them and those who are more able and confident may make a table and work more systematically. All of these methods may provide the correct answers, but obviously some will take longer than others.
- In the early stages of using apparatus in a new way, recording may take the form of drawings or words and drawings. Some children will gradually find this time-consuming and will simplify their recording independently. Others may need your suggestions and encouragement. As a teacher, you will need to work out carefully when a child is ready to use a standard mathematical symbol or format, so that recording is based on full comprehension. It is crucial that you do not force the children into formal and standard methods of recording calculations before they have fully grasped the process and are confident in the methods.

Exploring and investigating

Traditionally, primary mathematics has almost always tended towards short, directed tasks that result in 'right' or 'wrong' answers. The activities in this course provide a balance between short, fairly self-contained activities and open-ended investigations that may sometimes require the children to collect information, make their own observations, ask questions or complete activities at home. Most of the activities are designed to develop children's awareness of the range of mathematical possibilities open to them when they are tackling a mathematical task.

It is important for the children to experience a range of different kinds of questions and tasks, and to develop the understanding that mathematics gives us tools to investigate and test out our answers to real-life questions. In most cases, there is more than one way to solve a question. Take a simple question, such as 10 – 8. This could be solved in a range of ways: by working out the difference between 10 and 8, counting on from 8 to 10, using a number line or a chart, breaking the 10 into two 5s and the 8 into a 5 and a 3, and so on.

As much as possible, allow the children to take control, make decisions and explore the many avenues that can arise from a simple starting point. Always encourage children to ask *What if … ?* and *Why?* when investigating. These questions may lead to new challenges, fresh understanding and the development of new skills.

Many investigations have no final 'answer' or easily accessible generalisation for the children. Some have a simple pattern or rule that may be discovered and explained. However, many children will want to know why certain patterns repeat, and offer explanations about the rules that govern them. This is the first step towards generalisation, and you should encourage this by asking questions, for example: *Why is the same number added each time?* or *Can you guess what will happen next?*

The value in investigations is in children following them as far as they can, and in the new skills that they acquire on the way. For some children, the early, often concrete, experimentation is enough to give them confidence, and increase their enjoyment of using already acquired skills.

Use investigations to introduce new concepts. For example, you can introduce number patterns by presenting number chains (sequences), and letting the children work out what comes next or what is missing. Introduce geometric patterns through explorations of colour arrangements using pattern blocks, tiles or geoboards. The children can explore the relationship between 2D shapes and 3D shapes by exploring and identifying the 2D shapes that are the faces of the 3D shapes.

As the children develop an investigative approach, help them to work in a step-by-step fashion, and remind them of the question they are working to answer. Making new observations and discoveries along the way is encouraged, but ensure these do not eventually distract from the question they are working on. Ask questions such as: *What does this show us? What do you notice? What do you wonder? How can you use this to work out the answer? What other information do you need? How can you find it?*

Many everyday objects can provide rich sources of investigative work. The 100 square, addition square and multiplication square all contain many fascinating patterns. Children can also explore patterns in shapes and solids, such as the relationships between sides, corners and angles in 2D shapes and between faces, edges and vertices in 3D shapes.

Number talks

Number talks are open-ended, meaning they present a question that has multiple possible answers, and multiple possible strategies for solving.

Children can use agreed hand signals as shown below, instead of calling out or putting their hands up. However, these may not be culturally appropriate in all countries, and you may want to come up with alternative signals that are suitable for your class.

- **Step 1: Present the problem and give time for independent thinking**
  First present the problem, then ask the children to think to themselves quietly how they could solve it. They can use hand signals (thumb up means 'I have a way to solve it') to show when they have an idea.

I have an idea, and one way to solve it

Note that when children signal that they have an idea, you can encourage them to think of more ways to solve it. They can indicate the number of ways by raising more fingers, as shown.

I have two ways to do it

- **Step 2: Share in pairs**
  In pairs, children share their suggestions for how to work a problem out. Note that explaining how they solved it is as important as (or more important than) the final answer. They can use a hand signal to indicate they are finished, such as a fist to the chest.

I have finished.

- **Step 3: Children share their strategies**
  Ask individual children to share their strategies with the class. Discuss each strategy as the child demonstrates for the class. Continue using hand signals so that children can engage with the discussion and take turns to share their own working. The children can use thumb and little finger out for 'I agree' or 'Me too'; holding their hands with palms down and moving them in a criss-crossing motion for 'I respectfully disagree' and raised finger for 'I have another way to do it'.

I agree or Me too

- **Step 4: Discussion**
  The discussion will take place as other children show that they agree or disagree, and offer alternative strategies.

I respectfully disagree

Principles of the number talk strategy

The number talk strategy aims to get all children to engage actively and creatively with mathematics. To achieve this engagement, it is important to remember the 'big ideas' behind the number talk:

- *Value the contributions of all children.* Throughout the discussion help them to articulate their thoughts and reach a clearer understanding of how they are working things out. The emphasis is on the process and understanding, not on reaching the right answer fast.

- Children often answer in a roundabout way, with a very general idea or 'sense' of what they want to say. Use questions to *clarify, refine and explore* the meaning behind each response.
- Encourage children to explore *conceptual explanations*, rather than procedures.
- Gently *explore interesting mistakes* because these provide opportunities for deeper understanding. We can often better understand how something does work by reaching a clearer understanding of how it doesn't work.
- Praise children for effort, for making sense of a problem and for *persistence in the face of difficulty*, rather than for efficiency. We like to tell children that if it is difficult, that tells us we are in the place of learning. If it is easy, that means we've already learnt it and it's time to find more challenge.
- Create a *safe community space* where children feel they can share their ideas without criticism or judgement.
- Encourage *sharing ideas*, rather than competition. Children should understand that learning comes from connecting with other people's ideas and building on them.

Finding out more about number talks

It is important to remember that number talk is a format or technique, not a specific lesson or set content.

- View number talks in action, shared by teachers, on the Internet.
- Find posters and strategies for number talks through an online image search.
- Books on number talks include: *Number Talks: Whole Number Computation, Grades K-5: A Multimedia Professional Learning Resource* by Sherry Parrish; *Classroom-Ready Number Talks for Kindergarten, First and Second Grade Teachers* by Nancy Hughes.

Numberless word problems

Many teachers fear problem solving because they present a problem, and then watch as children try to guess which operation to use, or which numbers to arrange into a number sentence. Numberless word problems offer a useful strategy for encouraging children to think and understand questions before looking for the 'maths answers'.

Here is an example. Say you want children to practise their addition. Instead of giving a problem in traditional form, we start with a numberless form of the same problem, and work very gradually towards presenting the numbers.

> *There are some children in the playground. Some more children come out to play. Now there are many children in the playground.*

The numberless word problem does two things:
- It removes the numbers.
- It removes the question.

This forces children to slow down and think about the problem before they can attempt to solve it.

To present a numberless word problem, follow these steps:
- **Step 1:** Present the word problem you want to use, but remove the numbers and question. So, reword it using words such as *some, more, joined, went away, took away, fewer*. Ask the children to tell you what is happening in the story.
- **Step 2:** Ask children to think about what the question could be in this story. You will need to give plenty of thinking time, and invite the children to be 'question detectives', guessing from the clues what we are going to ask. They may have a variety of suggestions, such as: *How many were in the playground to start with? How many children came out? How many were there altogether?* Once they have identified the possible questions, you can reveal the question, but still without numbers.

> *There are some children in the playground. Some more children come out to play. Now there are many children in the playground.*
> *How many children came out to play?*

- **Step 3:** Have the children identify what information they need in order to solve the problem. First, ask the children if they can answer the question. They cannot – because they don't have the information they need. Let them suggest what they need to find out in order to solve it. Give them the first piece they need, by crossing out *some* and putting a number in:

> *There are ~~some~~ 15 children in the playground. Some more children come out to play. Now there are many children in the playground.*
> *How many children came out to play?*

Again, ask whether they can answer the question. When they say no, let them work out what other information they need to solve it. Let them work out what else they need to know. Finally, cross out *many* and change it to a number (24).
- **Step 4:** As with number talks, give the children time to solve the problem using their own strategies. Follow the steps of a number talk as you let them share and discuss their strategies.

Some children may struggle to understand the relevance of mathematics in their everyday lives. This course places great emphasis on making children aware of the relevance of mathematics in real life.

In this Teacher's Book, you will find ideas for using the child's own environment as a stimulus for mathematical activities. The Pupil Book and Workbook frequently require the children to look at the mathematics in the classroom, the playground and their own homes. Each set of activities and problems requires new skills and fresh understanding. Many questions are open-ended or have no exact solution, and the children are asked to make predictions, generalisations and estimates, and to evaluate their own answers. Encourage these skills in all areas of the curriculum. Children use their understanding of mathematics at home and at school, in situations such as sorting toys or other items, telling the time, looking for and making patterns, helping to prepare food, and playing board and other games.

In school

In school, there are many opportunities for you to teach mathematics through familiar situations, so that the children understand why it is useful and appreciate the order and sense that mathematics gives to life. For example, the children can identify the date each day, as well as the time at various points throughout the lesson. Registration, dinner money, timetables, sorting and putting away equipment will provide a range of relevant experience in data work, measures, shape and space as well as number.

In play

Children of all ages should have opportunities to play both in and out of school. This offers them the freedom to explore new situations, to make discoveries for themselves and to be creative. Unfamiliar mathematics equipment should be introduced through play, with the children exploring the functions and possibilities of the materials. A good example of this is a geoboard – you can show the children how to use it, and then let them work out what different kinds of shapes they can make using the elastics and the board.

Construction kits offer children the opportunity to explore shapes and inverse operations, through building and dismantling. Note that for any materials that are unavailable at your school or in your area, you may be able to find an online version.

At home

Part of the teacher's role is to involve parents and guardians in the children's learning and there are many opportunities for children to develop their mathematical understanding in simple ways at home. However, in order for this to happen, you may need to give parents some guidance about how easy this can be.

Many parents lack confidence in their own mathematics ability, and have the idea that maths is something complicated that children need to learn at school. A useful technique is to offer a 'learning café' – a short presentation for parents where you present simple ideas that they can use at home with their children. If time is an issue and parents cannot attend a session like this, adapt it to a page or booklet of easy home ideas for supporting mathematics. You can do this by pointing out the contexts in which we can use mathematics at home. Here are some examples:

- *Counting:* The children can practise counting through simple games and when using building blocks. Shopping is also a great opportunity for counting. Parents can ask their children to count out given numbers of items, or identify how many things are in a bag or box (how many eggs in differently sized trays, how many rolls in a bag, how many oranges in a container, and so on).
- *Patterns:* Show parents how to make patterns out of construction blocks. They can also talk to their children about everyday patterns that they notice in the world around them.
- *Shapes:* Explain to parents that identifying and sorting shapes and everyday objects is a key skill. Naming and identifying shapes around the house can help consolidate this skill.
- *Ordering:* When tidying and ordering the home, we use the skills of sorting things, arranging from biggest to smallest, identifying what belongs where. These are all skills that support mathematical reasoning. Encourage parents to involve their children in these activities.
- *Measures:* When we cook, we measure ingredients, either with scales or with cups and spoons. Parents can show their children how to use these.
- *Time:* Draw attention to telling the time, comparing how long things take, understanding minutes and hours as well as times of day, and so on.

You can find additional parent notes for each level on Oxford Owl. See page 9 for more details.

Section 3 An environment for exploration and play

All teachers have their own preferences about how best to organise the available space. However, here are some useful guidelines for any classroom.

Storage

Always store equipment so that the children have easy access to it and you can check it periodically. Label all items clearly and encourage the children to make their own decisions about what they need. From the very beginning, insist that the children pack up and return equipment to the correct storage containers or shelves.

A mathematics centre

Some teachers may choose to have a dedicated area, maybe a corner, of the classroom where they keep maths materials such as counting frames, ten frames and Numicon, prepared number game boards, geometric shapes, and so on. You can also display posters about shapes and numbers, patterns and other topics as these come out of the weekly lessons. Other teachers may prefer to integrate this into the rest of the classroom.

Recycled resources

Many household items can be repurposed to make mathematical equipment. Invite the children and their families to collect some of these materials at home in order to keep your classroom well resourced:

- bottle caps
- egg boxes
- clean glass jars
- buttons
- plastic caps from juice or milk cartons
- plastic yogurt pots
- beads
- natural objects such as shells, seed pods or stones for the children to arrange in size order.

Store the materials neatly in plastic containers or jars.

Individual work, classwork and groups

Whole class teaching

Often you will work with your whole class, especially at the beginning and end of a lesson. The course offers plenty of ideas for this kind of approach. The children may be seated together on chairs or on the floor, arranged in a circle for a discussion, or clustered around you for a demonstration. If you are calling on individuals to show things to the group, make sure there is enough space at the front for them to come up, and make sure everyone can see and hear the volunteer child's answer.

Set up some classroom rules at the beginning of the year and go over these at the beginning of each term or even each week if necessary. Classroom rules might include listening when others are speaking and waiting turns. Many teachers find it useful to use signs for *agree*, *disagree* and *I know the answer* rather than having the children raise their hands or shout out. The traditional approach of raising hands can be intimidating for quieter children or those who take a little longer to think about their answers. Hand signals allow teachers to give more time to those in the group who need it. The illustrations on page 17 show some commonly used hand signals.

Group work

You can group the children in similar or mixed-ability groups, to suit the purpose of the work. This gives the children opportunities to collaborate and to discuss their work with each other and with you. It allows peer teaching to take place and for the work to be matched to their needs. It also allows you to work simultaneously with a number of children and this minimises the need for repeated explanations to individuals. Group teaching is an effective form of classroom organisation for both teacher and children.

Working individually or in pairs

At times it may be appropriate for the children to work as individuals or in pairs, to provide extra help to children who need it, or to stimulate and challenge children who have grasped a concept quickly. Working individually gives the children the opportunity to concentrate on their own thinking, to develop this through investigations and problem solving, and to experiment with materials. Children working in pairs have the opportunity to develop collaborative skills, to play games together and to share ideas in an investigation.

Using manipulatives

Manipulatives give children an opportunity to experience the use of mathematics through concrete materials and objects. There are suggestions for different kinds of manipulatives in the unit-by-unit teaching guidance section of this book.

Many of the practical activities recommended for kindergarten level may still come in useful in the Level 1 and Level 2 classroom, particularly if many children in your class did not attend a kindergarten class. For this reason, we are including the following activities here for teachers that may need to bridge gaps from the kindergarten level.

Counting apparatus

Interlocking cubes

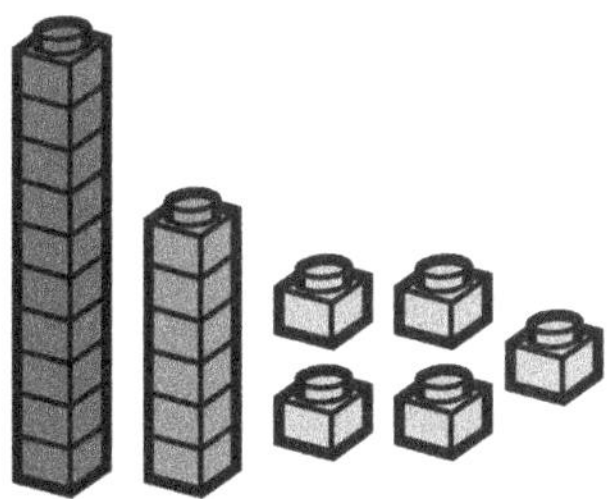

Interlocking or 'snap' cubes

Interlocking cubes are an invaluable resource for learning to count, especially as the children learn their number bonds up to 10, and then begin to work with tens and ones as they work with numbers up to 100.

Rekenreks

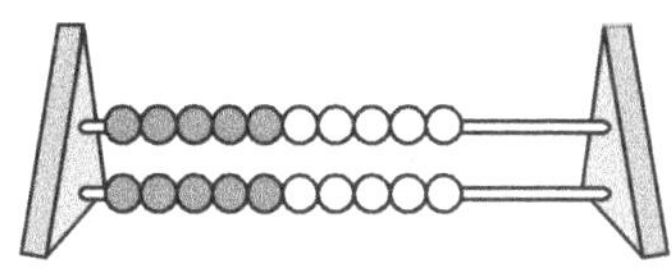

A rekenrek

The rekenrek was designed by a Dutch mathematics researcher to support children in developing their number sense in a natural, intuitive way. *Rekenrek* means counting rack or counting frame. At the primary level, the device consists of *two horizontal bars, each with ten beads*. The *beads are in two sets of five*, shown by different colours (usually red and white). The rekenrek looks a little like an abacus, but the bars do not represent place values.

When we work with a rekenrek, we start with all the beads pushed over to the right, in the 'resting' position. To count or represent a number, we move beads over to the left. The rekenrek can be used for a range of skills, including:

- developing subitising skills within 5 and within 10
- counting to 10
- addition and subtraction
- composing and decomposing numbers up to 20
- recognising doubles and near doubles
- skip counting.

Working with a rekenrek allows children to develop a range of strategies intuitively.

If your school does not have access to commercially available rekenreks, you may be able to make some simple ones with dowel sticks and beads. You can also search for an online interactive rekenrek app.

Ten frames

A ten frame is a 2×5 rectangular grid, used most commonly with counters of one or more colours.

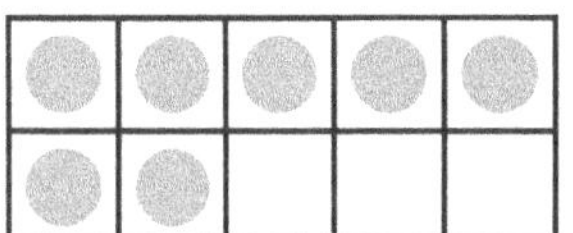

Using a ten frame to represent 7

Ten frames help children build and visualise numbers to 5, 10, 20, and 100. They can use the frames to count, represent, compare and calculate, working within a specific number range.

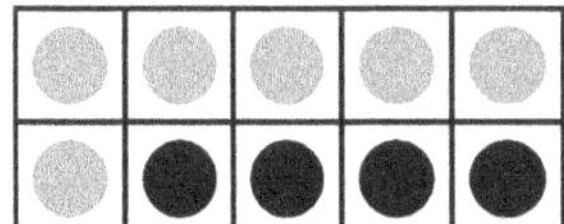

Using a ten frame to represent 6 + 4 = 10

Ten frames help children to see equivalences between quantities. So, for example, they will quickly see that 10 is equivalent to two 5s, and that 20 is equivalent to two 10s, or to four 5s. This helps them move away from linear, one-by-one counting and develop a range of counting strategies.

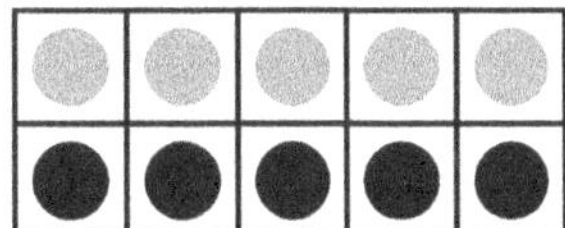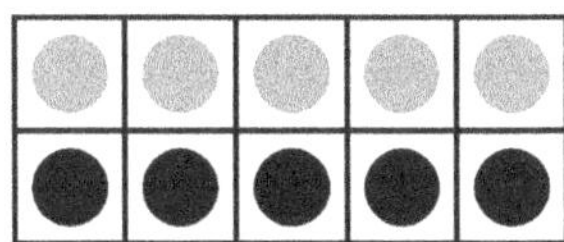

Using ten frames to represent 20 as two tens or four fives

Your school may have access to wooden or plastic physical frames. If you don't have these at your school, you can easily make your own ten frames and use sets of counters (five of one colour and five of another).

You can also search for interactive ten frames online.

Numicon (Number frames)

Numicon

Numicon is another powerful and versatile concrete resource that supports children to visualise and work with numbers. Each hole in a shape represents a number.

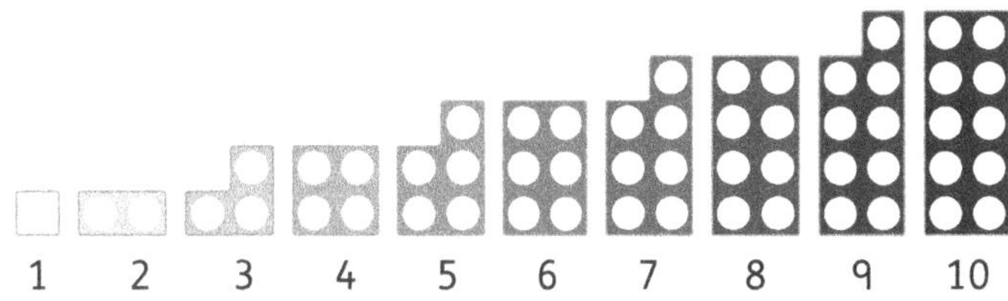

As with ten frames, children can also use the Numicon shapes to count, represent, compare and calculate. When thinking about 5, for example, they can quickly see that 5 is 1 less than 6, but 1 more than 4. Using a 3- and a 6-shape and comparing it to a 9-shape they can see that 6 and 3 make 9.

Using Numicon to represent 6 + 3 = 9

For more information on using Numicon to teach mathematics, and the full range of Numicon products, visit:

https://global.oup.com/education/content/primary/series/numicon

Base-ten blocks

Base-ten blocks are a concrete resource made up of blocks of varying sizes. Each block can represent a different value. For whole numbers, they represent:

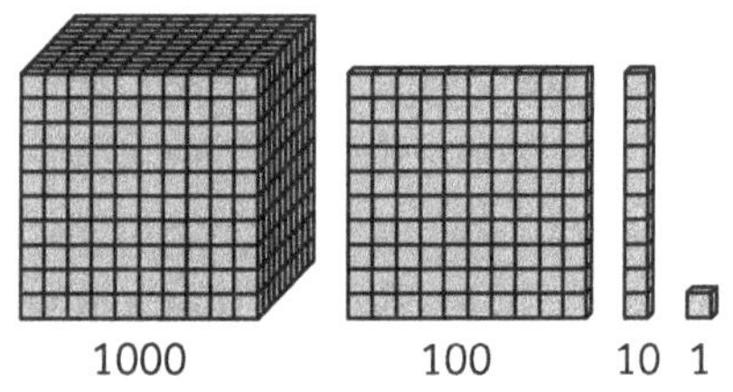

They can also be used to introduce decimals, representing, ones, tenths, hundredths and thousandths respectively.

Like ten frames and Numicon, base-ten blocks help children to represent numbers, compare and calculate. They are particularly effective when used to support partitioning, for example into tens and ones.

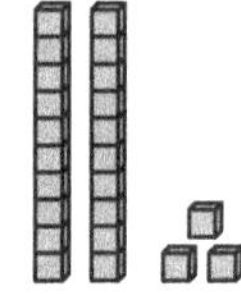

Representing 23 as two tens and three ones using two 10-rods and three ones cubes

Using base-ten blocks and place-value charts helps children to see how numbers change when multiplied and divided by powers of 10. Subtracting using base-ten blocks is a great, 'hands-on' way to introduce and support exchanging.

Bar models

Drawing bar models is excellent problem-solving strategy that children can use in both primary and secondary school. A problem that may otherwise seem very challenging can become more accessible when you 'draw' it.

Children should be encouraged to interpret bar models as well as draw their own when solving word problems.

There are two main types of models: the part-part-whole bar model and the comparison bar model. Which you choose to use will depend on the problem you are trying to solve.

For example, Tyra has 32 and she give 13 to Esme. How many marbles does she have left?

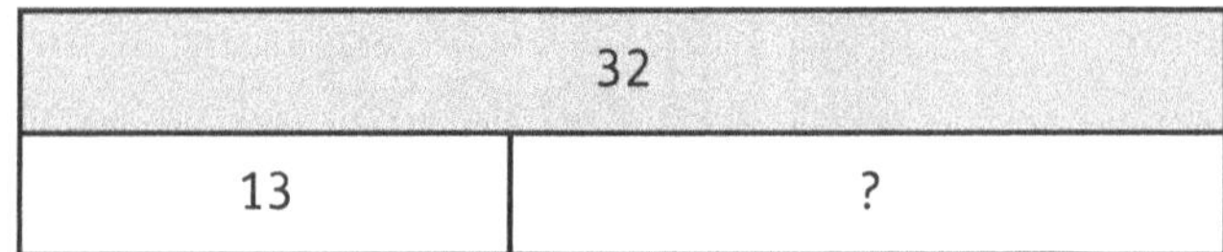

A part-part-whole bar model

Lewis completed 142 laps of the running track in one week.

Joshua completed 52 laps in the same week.

How many fewer laps did Joshua complete than Lewis?

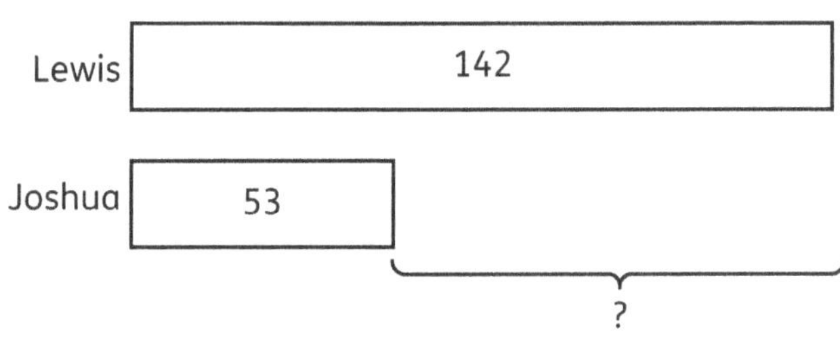

A comparison bar model

Place-value cards

Place-value cards have an 'arrow' or point on the right-hand side. Children can organise the cards horizontally or vertically to represent numbers in expanded notation. They can overlap cards and line up the arrows to form multi-digit numbers.

If any children have not previously worked with place-value cards, you will need to teach them how to use them. Begin by pointing out the arrows on the cards. Explain that these arrows always go on top of each other when you are making a number.

Place-value cards are an important teaching and learning resource and it would be useful to have a set available for each child. If possible, laminate the cards to make them more durable. (If you are making a set for each child, you may like to send the cards home for parents or carers to cut out.)

In Level 2, the children only need to work with numbers to 100, so you only need to prepare tens and ones cards. At higher levels these can be extended to as many places as needed and also to the right to show decimal places.

Here is a basic set of tens and ones place-value cards:

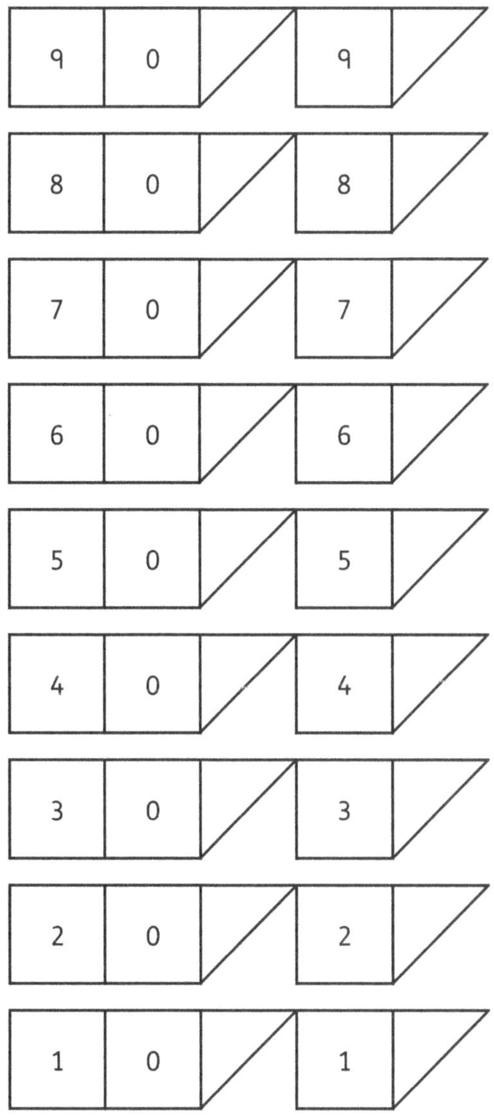

Place-value tables

Children can use place-value tables to build numbers using counters or base-ten blocks. If possible, make place-value tables on card and laminate them to make them more durable.

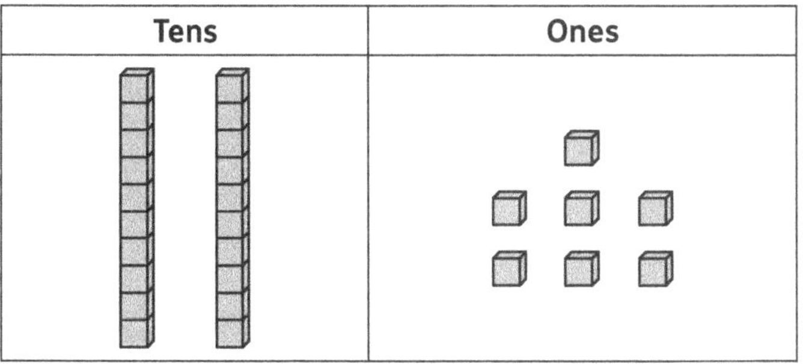

Tens	Ones

In Level 2, the children only need to work with numbers to 100, so place-value tables only need Tens and Ones headings. At higher levels these can be extended to as many places as needed and also to the right to show decimal places.

Materials for sorting and matching activities
Containers

Provide a range of containers of different sizes and shapes, and access to sand and/or water. Allow the children to fill and pour sand/water from one container to another. Ask questions and use the activities to develop the basic vocabulary of mass and capacity – *heavy, light, heavier, lighter, full, empty, nearly full, nearly empty, holds the same, holds a lot, holds a little, holds more than, holds less than*, and so on. These activities develop hand–eye coordination skills, as well as vocabulary, and also help the children to develop the idea of conservation of measure (for example, to learn that a wide flat container may hold the same as – or more than – a tall, thin container).

Feely bag

Prepare a 'feely bag' for the classroom. You will need a range of different items (for example, plastic cutlery, pens, rulers, buttons, combs, keys, coins, feathers) and a bag that you cannot see into. The children take turns to put their hand into the bag and select an item. They should hold the item and describe how it feels it to the class (*It is big, it is soft, it has smooth sides*). The others try to guess what it is. After some discussion, let the child remove the item and show it to the class. They should then describe it using its characteristics (including colour, size and shape). For example, the child might say: 'This is a comb. It is brown, and long and quite thin. It has teeth on one side and a flat handle.' You can include matching in this activity by drawing pictures of the items in the bag, or tracing around their outline. Ask the child to match the chosen item to the correct picture or outline.

Collections

Arrange the children in groups. Give each group half an egg box (or any other container with a number of compartments). Ask them to walk around the classroom or take them out into the school grounds and ask them to collect six items that are different from each other. When they have collected six items, the groups take turns to describe the items and say how they are different. For example: 'This is a big brown seed. It is hard and rough on the outside. This is a small green leaf. It is smooth and round.' You can vary this by giving different instructions, for example: *Find six items that are all green, but all different in some way*.

Everyday objects

Matching pairs of everyday objects helps children to develop the idea of one-to-one correspondence. For example, ask the children to physically match items such as:

- straws to bottles or cups
- children to chairs
- bags or coats to hooks or lockers
- children to pencils or books
- buttons to buttonholes
- hats to heads
- socks to shoes.

Provide crayons in different colours and ask the children to match them to counters using colour as the matching characteristic. Use this activity to teach the names of the colours if necessary. You can vary this to match any items using colour.

> With all activities using colour, if you have colour-blind children in your class, adapt the activity to use patterns, or combinations of colours that are within their visual range.

Shape jigsaws

Prepare a set of 'shape jigsaws' for the classroom. For each 'jigsaw' you will need two sheets of card the same size. Divide the sheet up into shape pieces like this:

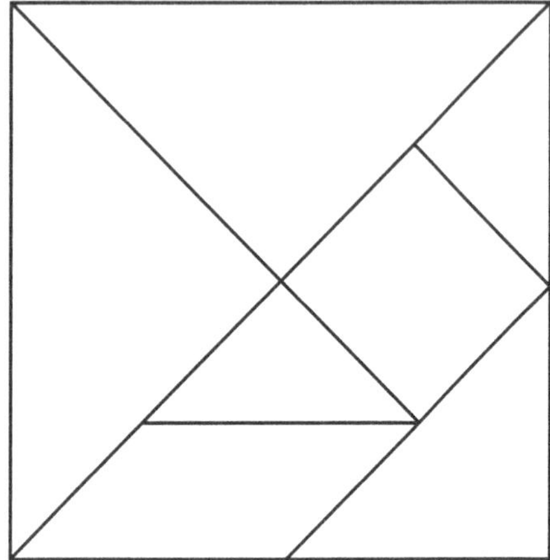

This shape jigsaw is based on the traditional tangram. However, you can vary the shapes to make different puzzle boards.

Draw the same shape lines on both sheets. Leave one sheet intact. This is the guide board for the children. Cut along the lines of the other sheet to make a set of shapes. Store each puzzle (the guide board and the set of shapes) in a zip-lock plastic bag or large envelope. Give the puzzle to individuals or pairs of children and let them make the jigsaw by placing the cut-out shapes on their matching outlines. Note that the names and attributes of the shapes are not the focus of this activity, rather the point is to encourage the children to match like-to-like shapes.

Paint strips

You can prepare a fun matching activity using the paint colour strips you get from hardware or paint shops. You will need two of each strip for the activity. Use, for example, a strip showing shades of green. Leave one strip intact (or cut out the different shades and paste them in a random order on another sheet of card). Cut the other strip into pieces so that each piece shows one of the shades of green. Store each puzzle in its own zip-lock bag or envelope with the colour written on the outside.

Arrange the children in small groups and let them play with the puzzles. They should display the fixed colour strip (or sheet) and have the other pieces face down. They then take turns to pick up a piece and match it to the correct colour on the game board. To extend the activity, ask the children to identify items in the environment that match each shade of green (for example, the leaves on the tree outside are a dark green like this strip, the paint on the classroom wall is a very light green like this strip).

Interlocking cubes

Prepare sets of interlocking cubes of different lengths (from one to about ten cubes). Cut lengths of wool or ribbon to match the length of the sets of cubes. Ask the children to match the lengths by comparing and then laying the wool next to the corresponding set of cubes. Extend this by including some pieces of wool that don't fit. Encourage the children to talk about why these don't fit. Say, for example: *This piece is too long for all the sets of cubes. This piece is too long for this one, but it is also too short for this one.*

Egg boxes

Prepare a matching activity using egg boxes. Draw or stick one to four stars in each compartment. Ask the children to sort beads or other small objects to match the number of stars (one-to-one correspondence). They do not need to know the numbers or be able to count to do this.

Numbers, dots, objects and cards

To develop number sense, include matching activities that involve numbers of items arranged in different ways. For example, prepare cards with one to five items (shapes, dots, squiggles, fruits). Ask the children to make pairs of cards with the same number of items. Once you have taught numerals and number names, you can extend this activity to matching the amounts and then matching those to numerals and sorting them in order from fewest to most or vice versa.

Collect pictures of numerals (from 1 to 10 only) in different contexts and styles from magazines, greeting cards, calendars and product labels. Stick the individual numerals onto card and keep them in a box. Give the children a handful of cards and ask them to sort them into groups of symbols that are the same. They do not need to know the number names to do this, they can simply compare the shape and form of the numerals to match and sort them.

Games

A game is different from a teaching activity because there is an element of luck and chance in a game, sometimes leading to a winner. Some games are played alone, and children try to improve on their own results, while other games are played in pairs or groups.

Make sure the children understand the rules of the game before they start playing. If they don't, they may not play the game properly and this can result in conflict with others. The children should learn to:

- take turns and wait for their turn
- move their own game pieces only
- help each other if they can.

The children also need to learn that in games involving luck, players don't win or lose on the basis of ability. Explain that those players who get the best luck in a particular game win; those who lose were just less lucky.

Winning graciously, as well as losing and playing fairly, are important life skills.

When the children play games in class, do not keep a record of who wins and who loses. If the children like a game, encourage them to play it in their free time and to take it home and play with family members if possible.

Section 4 Activity bank of warm-ups and mental maths

This activity bank includes a range of mental maths activities, as well as some suggestions for support and consolidation activities. Most of the ideas are for number work and operations, although there are also some suggestions for the other strands. You can use this as a resource for activity ideas for your class. The unit-by-unit lesson plans refer back to specific activities in this section as unit openers, suggestions for extra support, and consolidation.

Try to include 10 minutes of mental maths activity each day. This section includes many examples that you can use as they are, or adapt to suit your own classroom. It provides a range of different types of activities (factual recall, games, grids, tables, problem solving and puzzles) to show some of the ways in which you can approach the mental maths part of the lesson. However, this is not a definitive list and some activities will appeal more to some classes and teachers than others. If you need additional ideas and suggestions, search for mental maths warm-ups online.

> Avoid any games that use countdown clocks or timers which turn mental maths into a race or speed test. Speed testing can be damaging to children's mathematical development and cause maths anxiety.

Number work and operations

Physical counting warm-ups
- **Skipping rope counting:** Bring a skipping rope to class. Give individuals a chance to try to skip continuously to the highest number. The whole class chants the count as they go. Make sure as many children as possible get a turn.
- **Ball throw counting:** The children stand in a circle and throw the ball across the circle to a new person each time. How many throws can they keep it going for before the ball drops?

- **Focus count:** This is a quieter activity. The children stand in a circle. One child says 'one', another says 'two', another says 'three', and so on. The speakers should go in a random order (not around the circle in the order in which they are standing). So, children need to 'sense' when it's their turn. The challenge of the game is that two children must not speak over each other or at the same time – if this happens, the counting starts from zero. This game is fun counting in ones or skip counting in any multiples.
- **Walk and group:** The children walk around the classroom space. When the teacher calls out a number, they get into groups of that number. Count up the groups, and how many are over. Ask the children to check your totals against the known total of children in the class. This game is also useful as a warm-up for division with remainders.

Number names and numerals
Early in the year you could use the mental maths slot to reinforce and revise number names and numerals 1 to 100 using flashcards. One side of each flashcard can show the number name in words, the reverse side can show the number in numerals.

Show a series of number names on them, such as *eighteen* or *thirty-five*. Display each card for a few seconds and have the children write the number in numerals. You can either let the class take turns to say the numbers aloud or you can do this activity in silence, to test that the children can read the number names and relate them to their numeric equivalents. Check the answers as a class by turning the cards over to show the correct numerals.

You can also use the cards to test number names by showing the numeral and having the children take turns to say the number you are displaying.

Order numbers and find one more/one less
As the children become more familiar with counting and numbers to 100, you can vary the work with 1–100 flashcards to include ordering and finding numbers

greater or less than a number as well as numbers between the given numbers. Say, for example:

- *This is the number 23. Write the number that is one more than this number.* (24)
- *I'm going to show you a number. Write down the number that is one less than this number.*
- *Write the number that is 10 more/10 less than the number I am showing you.*
- *Here are two numbers. Which is greater?*
- *Here are three numbers. Which is smallest?*
- *Here are two numbers. Write a number that is between these two numbers.*

Rekenrek counting and operations

Have children use rekenreks (see page 21) to model addition problems and subtraction problems.

Missing numbers

Display a number track of 2-digit numbers with some numbers missing. Point to an empty square and let the children guess the number that goes in it. Repeat for each missing number. For example:

24	25			28	

39		41			44

63		66			

To make it easier, have only one or two numbers missing.

To make it harder, leave out more numbers. Leaving out the first number means children have to work out both more than and less than a given number.

You can also vary this to use tracks with skip counting sequences.

Count backwards

Display a 2-digit number. Ask the children to read it, identify how many tens and ones it has and then count on and back in steps of ones or tens from the number.

Number facts

Play a game in which the children have to take turns to make up a 'fact' about the number. For example, using the number 55, they could say things such as:

- 'It is 1 more than 54.'
- 'It is 10 more than 45.'
- 'It is greater than 50.'
- 'It is between 50 and 60.'
- 'It is odd.'
- 'It is a multiple of 5.'

As the children learn more about numbers they will be able to give more complex facts, but at the beginning of the year aim to get three or four simple facts per number.

Show me a number

Give the children sets of place-value cards (see page 23). Ask them to sort their cards into ones and tens. When they have done this, ask them to show you some numbers starting with numbers that only use one card. For example, say: *Show me 3* (or 6, 7, 10, 40, 60, etc.).

Next, show the children how to put the cards together to build numbers. You may need to demonstrate. For example (putting the 10 and 5 together), say: *This is how we make 15.*

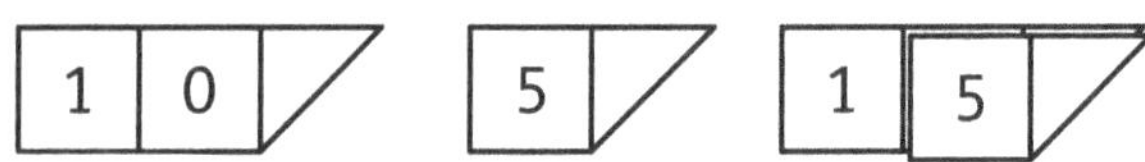

Check that the children can build different numbers by calling out some numbers and having them show you. As the children build numbers, they will begin to make connections and observations. They may notice that building the numbers is the same as adding numbers, for example 10 + 5 = 15. This is an important observation that forms the basis for partitioning numbers and written methods at later stages. Encourage the children to share their observations with the class.

Digit positions

Use place-value cards (see pages 22–23) and spend some time building numbers with the same digits in different positions, for example 39 and 93, and 17 and 71, to make sure the children understand that the position of the digit is important. Discuss the value of the digits in the numbers that the children build. Do this by building a 2-digit number (for example, 53) using place-value cards and ask the children to say what the 5 and the 3 represent (50 and 3).

Ask the children to make as many numbers as possible with a 1, 2 or 3 in any place. (They will need to work in groups and combine their cards to build these, or they will need to record as they make each number to keep track.) Repeat with different digits.

Place-value games

Play some games with the place-value cards (see pages 22–23) to challenge the children to listen and think carefully. Say, for example:

- *Make a number equivalent to 4 tens and 3 ones/ 4 tens and no ones/no tens and 9 ones.* (43, 40 and 9)
- *Build 45 and 54. Which is smaller/greater?* (45 is smaller and 54 is greater)
- *Build 19 and 21. Which is smaller/greater?* (Vary the order in which you give the greater and smaller numbers.)
- *Build a number between 33 and 36.* (Children build 34 or 35.)
- *Build an odd (or even) number between 11 and 19.* (Odd: 13, 15, 17; Even: 12, 14, 16, 18)
- *Build three different numbers with digits that add up to ten (73, 64, and so on).*
- *How many numbers less than 100 can you build with 4 in the tens position?* (10)

- *I am a number between 40 and 60 with one 5. What number could I be?* (45, 50 or 55)
- *Build a number that reads the same from back to front.* (11, 22, 33, 44, and so on)
- *Build a number with a name that rhymes with* fine *(nine).* (Other rhyming numbers could rhyme with bun (one); you (two); me (three); door (four); alive (five); sticks (six); leaven (seven); late (eight).)

More or less

Use place-value cards (see pages 22–23) to reinforce counting activities. These activities are useful because they require the children to build numbers physically and then partition them to replace digits, which helps them make sense of calculations involving 2-digit numbers. Use these activities, for example:
- *Build a number that is 10 more than 37.* (47)
- *Build a number that is 10 less than 83.* (73)
- *Build a number that is 3 less than 44.* (41)
- *Build a number that is 5 less than 80.* (75)

Building numbers in place-value tables

Give each child a place-value table (see page 23).
- The children pick a digit card and make the number in the place-value table with counters or cubes.
- The children pick a digit card for tens and a digit card for ones and place them in the place-value table to make a number (for example, a 2 and 5 for 25).
- The children play a game where they take turns to spin a 1–6 spinner, or pick a digit card without looking. They write the number in one of the columns in the place-value table. For example, if they get 6, they may write it in the tens or ones place. Then they spin, or pick a card, again and write the number in the empty column. Change the aim of the game so that sometimes the winner is the child who makes the greatest number and other times it is the child who makes the smallest number. You can include 0 as a challenge.
- Give the children some possible digits for each place value and ask them to work out how many numbers are possible with the given values. The children should list all the numbers they can make, for example:
  - *The tens place can have: 2 or 5 and there are zero ones.* (They can make 20 or 50.)
  - *The tens place can have: 4, 5 or 6 and the ones place can have: 0, 1, 2, 3.* (They can make 40, 41, 42, 43, 50, 51, 52, 53, 60, 61, 62 and 63.)

Sort and order number cards

Make sets of five or six different 1- and 2-digit numbers on small cards (you will need enough for each group or pair of children). You can easily make 100 cards by enlarging and then cutting up the 100 square. Shuffle them and put them into envelopes for use. Distribute the sets of numbers and let the children arrange them in order (from greatest to smallest or vice versa). You can also use these cards to sort into odd and even numbers, or numbers greater than 20 and numbers less than 20, and so on.

Guess the number

Play either as a class or in groups. Let the children take turns to choose a 2-digit number and jot it down. The group then takes turns to ask questions to try to guess the number. The child who has the number may only answer 'yes' or 'no'.

Show it four ways

Give each child an 'I can show it four ways' worksheet:

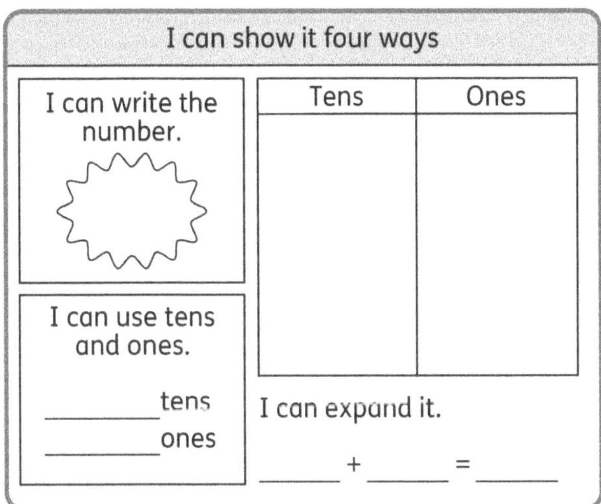

Give them a target number, for example 23, and ask them to show it in the four different ways on the table.

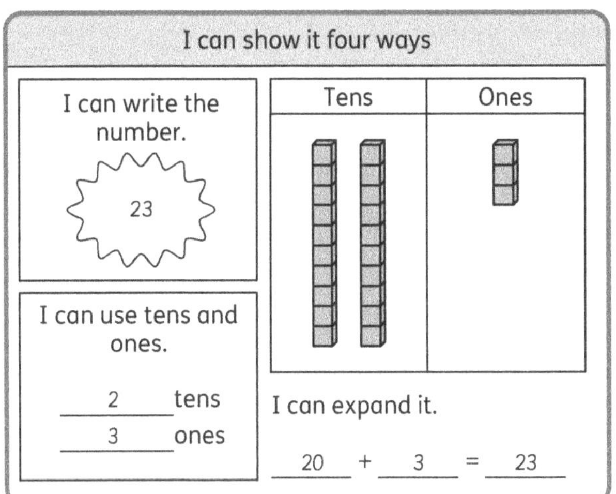

Four-part puzzles

Create puzzles that show numbers in four different ways: as a numeral (and number name), visually using interlocking cubes or base-ten blocks, as tens and ones, and as an addition sentence of a multiple and ten and a 1-digit number, as shown. Then cut these up and let the children make up each track with four equivalent representations of the number.

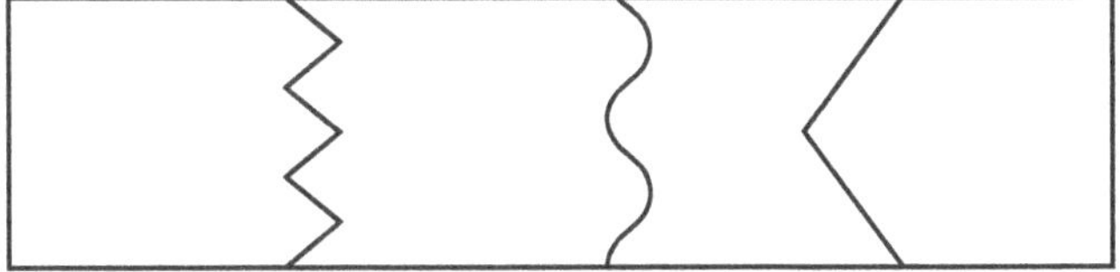

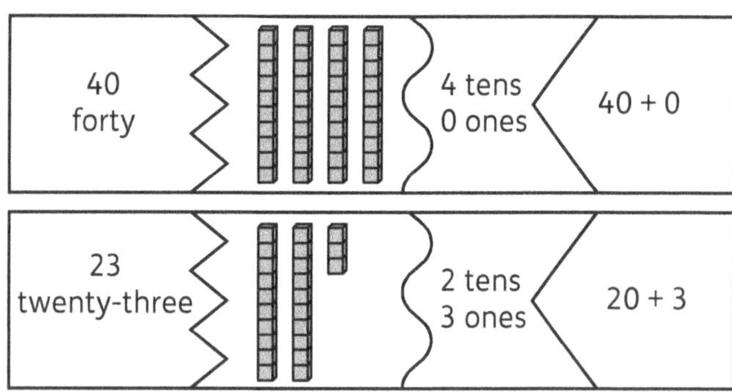

Compare using <, > or =

Ask the children to write down any 2-digit number. Write a random set of 2-digit numbers of your own on the board.

- Let the children make number sentences using your numbers and the number they have written down using the <, > or = signs.
- Let the children use mental strategies and jottings to find the sum of or difference between the numbers in their number sentences. Spend some time talking about the strategies they suggest.

Round numbers

- Ask the children to jot down a 2-digit number. You can specify that it should have two different digits or you can pick digit cards to generate random 2-digit numbers. Once they all have a number, make a 'human number line' with ten children standing in a row displaying the multiples of 10 (10, 20 … 100). Children then take turns to display and read out their own number and then hand it to the number line child with the closest multiple of 10. They can check each other as they go.
- Use any book. Turn to a page, for example page 33. Ask the children to write down whether you are closer to page 30 or 40. Repeat for a few different pages.

Find the rounding mistake

- Write several 2-digit numbers on the board. Round each to the nearest 10. Make sure some of the rounded values are incorrect.
  For example:

23 → 20	38 → 30
26 → 20	39 → 40
41 → 40	

Ask the children to find the incorrectly rounded numbers and to correct them. Repeat this activity using different numbers.

Rounding practice

Display a grid like this, where all numbers have a value other than 0 in the ones place.

56	75	49	19
45	51	19	15
59	82	81	87
13	62	91	14

You can ask the children to copy the grid, and have them rewrite the numbers, rounding them to the nearest 10 as they do so. Or you can work through the grid as a class, picking a child to round each number to the nearest 10.

Rounding challenge

Write a multiple of 10 on the board. Draw six arrows pointing to it. Challenge the children to write down six numbers that would round to this number.

For example, ask them to write six numbers that would round to 30.

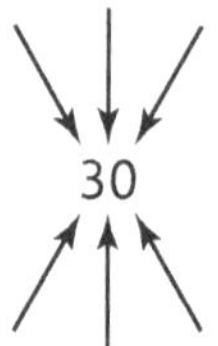

Round as you go

Ask the children to make a three-column table like this one:

10	20	30

Read out 8–10 numbers in the range from 5 to 39. As you say each number, the children should write it in the correct column to show the nearest 10. For example, if you say 15, the children should round it to 20 and write it in that column.

Estimate

Prepare an estimation jar. Use a small glass jar or plastic container and fill it with items such as beans. Stay in the number range 0–30 at this stage. Let the children jot down how many beans they think there are in the jar (to the nearest 10). Then show them a smaller container filled with the same items and tell them how many there are. For example, say: *This jar has 10 beans in it. How does that affect your estimate?* Discuss how their estimates would change if you used larger/smaller items to fill the jar. Discuss how you could check the estimates and then count the items in groups of tens or other appropriate groups.

You can also tell the children how many items a jar can hold when full, for example: *This jar can hold 80 marbles.*

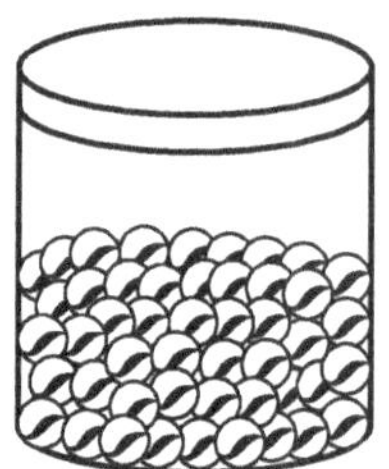

Tell the class that some marbles have been taken out and ask them to estimate how many are left, and/or how many were taken out. Discuss how they reached their estimates.

Estimate with dot cards
Prepare a set of cards with a number of large dots on them.

For example:

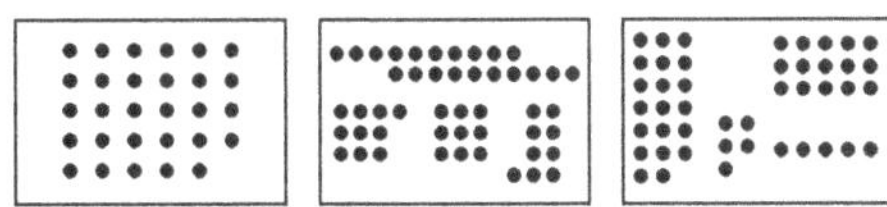

Display these and get the children to estimate how many dots there are on the cards. Spend some time discussing how they reached their estimate, for example, *the dots are in groups of about five, and there are six groups, so I estimated 30.* You may need to flash the cards for a few seconds and then remove them to avoid children trying to count. Remind them that you are asking them to *estimate*.

Estimate using 100 squares
Display a blank 100 square. Then display different 100 squares with different numbers of squares shaded and ask the children to estimate (to the nearest 10) how many squares are shaded and/or unshaded).

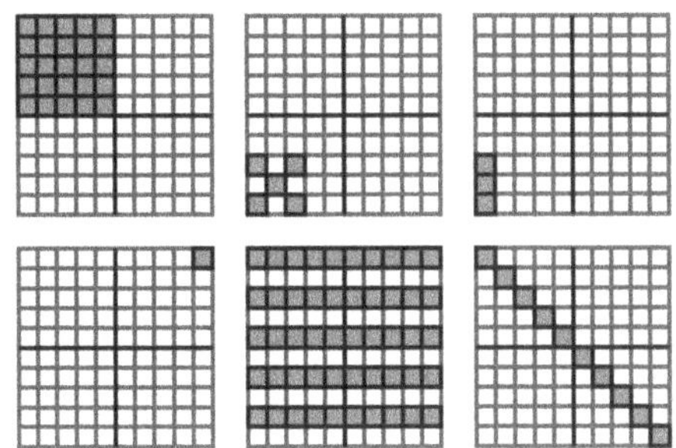

Discuss the strategies the children used to reach their estimates.

Estimate with number lines
Display a blank number line marked with 0 and 100.

0   100

Use an arrow (for example, drawn on a sticky note) to indicate the position of different numbers. Let the children estimate what the number could be and say how they decided. A child might say, for example: 'I think this is about 70 because it is more than half way between 0 and 100, but not quite in the middle between 50 and 100.'

What fraction is shaded?
Once the children have learnt simple fractions, you could show a range of shapes with parts shaded and ask them to estimate what fraction is shaded. Similarly, you could display a small set of objects with a ring around half or quarter of them and ask the children to say what fraction is encircled.

Target diagrams
Target diagrams can be used in different ways, for example to help with addition and multiplication.

Display some target diagrams, sticking moveable counters onto them and asking the children to work out the total score for each diagram (adding several small numbers). Repeat with the counters on different numbers. For example:

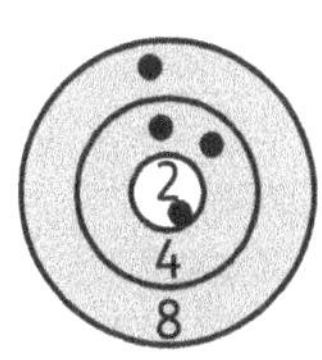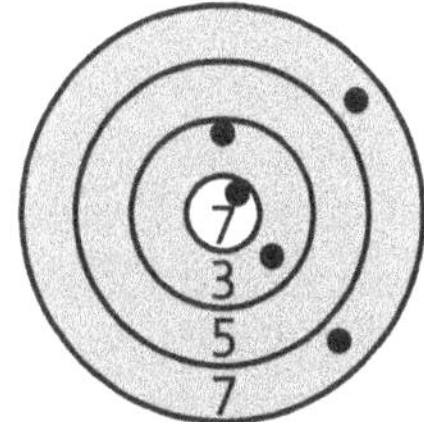

Prepare a target diagram like this one, with a 10 or 20 in the centre.

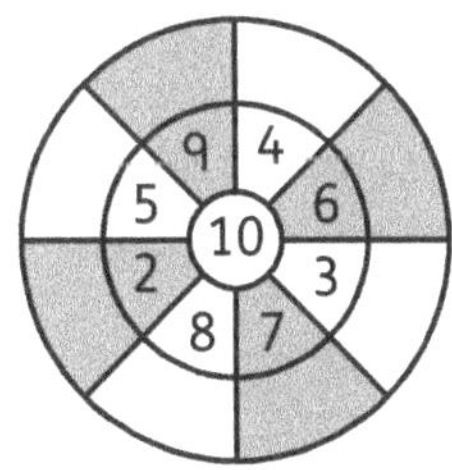

Tell the children that each sector adds up to 10 (the number in the centre) and ask the children to find the missing number on the outer ring. You can adapt this by writing a small number (1–5) in the middle and then saying that the outer ring minus the inner ring will give this result. Ask the children to find the missing numbers. Vary the task by sometimes providing the outer ring, sometimes providing the inner ring and sometimes providing some inner and some outer numbers.

You can also use target diagrams to help with multiplication. In the examples below, the children have to find the missing numbers on the diagrams:

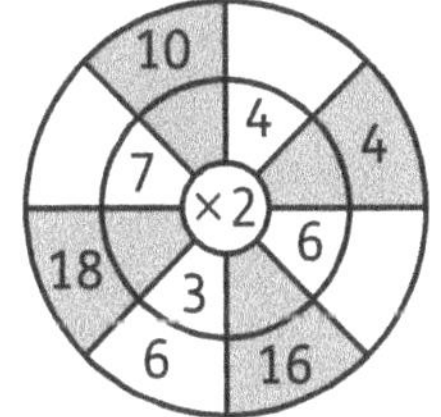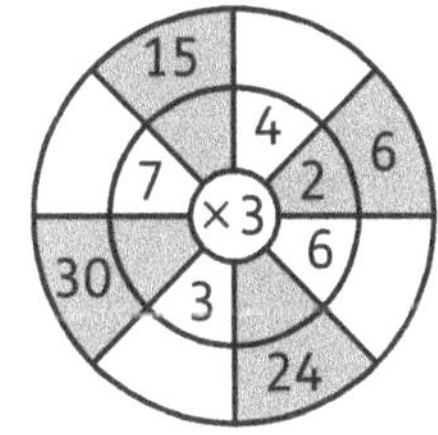

Target number calculations
Prepare a set of 'target' numbers, for example 14, 18 and 23. Ask the children to write as many calculations as they can to get to each number. You may want to limit this to focus on particular operations, for example: *Write as many addition sentences as you can that equal this answer.* Or you can leave it open ended and challenge the class to find as many different operations as possible.

Target number instructions

Test listening skills and computational fluency by reading instructions about a target number. The children can jot them down and they should record the answers and hold a thumb to their chest when they have them. Adapt the tasks to the skill levels of the class and what you have covered at that stage of the year. For example:

* Using easy numbers to start with, tell the children: *I'll say a number and you double it.* Then include an addition or subtraction, for example: *I'll say a number. You double it and then add (or subtract) 1 (or 10).* Then, adapt the task, for example: *I'll say a number, add 1 to it and then double it.*
* Make the calculation a chain calculation, for example: *Start with 2. Double it. Add 1. Then subtract 2. Add 3 and then find the number 1 less than that.*
* Give the answers and have the children make up a matching calculation. State the operation, for example: *In an addition, the answer is 9. Give me two numbers.* Increase the number range and the numbers, for example: *The answer is 12. Give me three numbers that add up to 12.* Change the target number regularly.

True or false/Explain the mistake

Use the 'true or false' strategy to test vocabulary and also to apply calculation skills. Give children a statement and have them say whether it is true or false. Discuss how they decided. For example, you could say:

* *1 multiplied by 7 is 8.* (False)
* *2 times 3 is 6.* (True)
* *The difference between 6 and 15 is 9.* (True)
* *The total of 3, 4 and 9 is 15.* (False)
* *15 taken away from 40 is 25.* (True)
* *Four lots of 8 are 48.* (False)
* *The product of 4 and 7 is 28.* (True)
* *50 divided by 10 is 3.* (False)
* *9 times 10 is 900.* (False)
* *45 divided by 5 is 9.* (True)

Another approach to this is 'explain the mistake'. Purposefully give children an 'interesting mistake', and ask them to have them work out what mistake the person made.

Use shapes to reinforce multiplication and division facts

Use the properties of shapes to reinforce multiplication and division facts. For example, display a triangle. Ask how many sides there are in 2, 5 and 10 triangles. Similarly, ask: *I have a number of triangles. There are 27 sides. How many triangles do I have?* Repeat for other shapes and numbers.

Number sequences

Give the children a rule for generating a sequence. Demonstrate with an example, such as:
The rule is add 10 and start on 15. The first three numbers are: 15, 25, 35 …

Ask the children to write down the next five numbers.

Give the rule and the starting number, and let them work out the numbers in the pattern, for example:

* *The rule is subtract 1. Start on 65.* (64, 63, 62, 61, and so on)
* *The rule is double the number. Start on 1.* (2, 4, 8, 16, and so on)

You can adapt this by giving the children sequences and letting them identify the rules, for example:

* *35, 30, 25 …* (subtract 5)
* *55, 65, 75 …* (add 10)
* *45, 40, 35 …* (subtract 5)
* *58, 61, 64 …* (add 3)
* *2, 4, 8, 16 …* (double or times 2).

Function machines

Prepare some function machine charts and let the children complete them. For example:

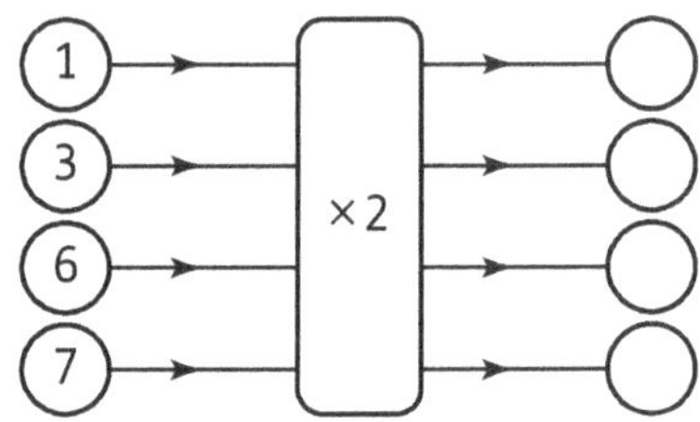

Measure

Show and tell the time

Ask the children to show or tell the time using an analogue clockface with moving hands, for example:

* half an hour before 2 o'clock
* half an hour after 1:00
* half an hour before 5.30
* half an hour after 4 o'clock.

Questions about measures

Ask which item weighs the most in each pair:

* an apple or a brick
* a pencil or an apple
* a kilogram of apples or your shoe
* a litre of cola or a glass of water
* a basketball or a tennis ball
* an empty cup or a full cup.

You can ask similar questions to compare lengths and capacities.

Estimating measures

Ask children to write estimates of the lengths of various familiar items in appropriate units, for example a shoe, a book, a desk, a pin, a pencil, the door of the classroom. Discuss what a reasonable estimate might be in each case.

Identify 2D and 3D shapes

- Display a number of 2D shapes. Give each child a set of flashcards with the names of the shapes. Show each 2D shape for a few seconds and ask the children to select and show the name of the shape. Vary this by saying the name and having the children select the correct shape from a selection of plastic or cardboard shapes.
- Say the name of a 2D shape and ask the children to write down how many sides it has.
- Show the class the name of a 3D shape, for example *cuboid*. Let them tell you what real objects are that shape. See how many you can list on the board.
- Use the mental warm-up session as an opportunity to explore 2D and 3D shapes used in buildings. Display pictures of homes, places of worship and/or murals and decorative patterns and spend some time identifying and naming the shapes used, symmetries and other properties.
- Show 2D representations of 3D shapes, show the children real items (for example, a cube, a soccer ball, a cereal box) and let them name these, and then find pictures of the 3D shape to make the connection between the real item and its representation.
- Display 2D shapes (or 3D shapes or objects) labelled A to F. Display the correct names of the shapes in a mixed order and ask the children to match them up.
- Use plastic shapes, or prepare and laminate cards with coloured shapes on them. Distribute these to the class so each child has at least two different shapes. Play a game in which you call out commands, such as: *Show me a red square*. All the children with that shape should then hold it up. Repeat for different shapes, moving quickly to make the children react quickly. If they hold up a wrong shape or colour, they have to put it down. Once they've lost their shapes, they can't play any more.

Shape jigsaws

Give each child a shape jigsaw based on a tangram puzzle (see page 24).

- Ask them to sort the shapes in the puzzle, and explain their sorting (for example, by shape, by colour, by size).
- Challenge them to make pictures using the puzzle's shapes, for example a house, boat or tree.

Pattern sequences

- Display shape patterns and have the children draw, show or describe the next shape. For example:

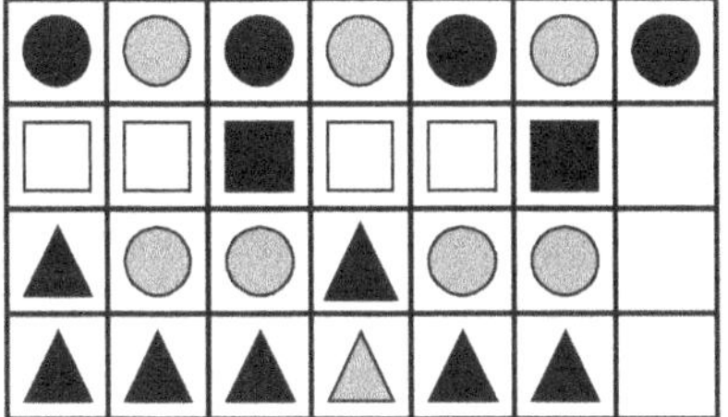

Encourage the children to tell each other how they decided what the next shape would be.

You can also display a set of shapes on a grid. For example:

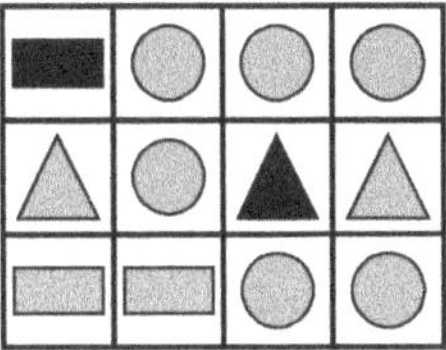

Ask questions such as:
- *Which shape is in the top right corner?*
- *What shape is above the red triangle?*
- *Which shape is left of the green triangle?*
- *Which shape is below the blue circle in the second row?*

Unit 1 is different from all other units in the course: it is a short introductory unit, and it serves to set up important norms in your class. This unit aims to prepare both children and teachers for the work ahead, and you should be able to complete it in one or two lessons.

As maths teachers, it is important to examine our own ideas about who can do maths and how mathematical ability and learning work. From the earliest years, we need to find real, experiential ways for children to understand that:

- asking questions is the best way to learn
- making mistakes is a key part of brain growth and learning
- everyone has the potential to learn maths – there is no such thing as a special 'maths brain'.

If we simply tell children these things, they will not believe us. We need to show them evidence of the ways that the brain grows and learns; and we should also show them that maths can include everyone.

Learning objectives

- Establish a classroom environment conducive to thinking and working mathematically.
- Set up positive norms.
- Establish key messages of growth-mindset mathematics.

Key words

brain   synapse   mistake   question

Unit introduction

Teaching guidance

- Write the word *maths* on the board. Ask the children to tell you any words or feelings that come to mind when they hear this word. Write all their ideas on the board.
- If children are shy, encourage a range of answers. For example, if children only mention maths topics (numbers, counting, adding), encourage them to suggest feelings, colours and other associations and

beliefs. They may also share fears and funny ideas about maths.

- Encourage them to think in many different directions. Possible prompt questions include: *What feelings do you get when you think about maths? What colours do you think of when you think of maths? When do you use maths in real life?*
- Summarise the ideas by pointing out that maths can mean different things to different people. Tell the class that they are going to explore some of the ways that we can all have fun with maths and learn well.

How I feel about maths

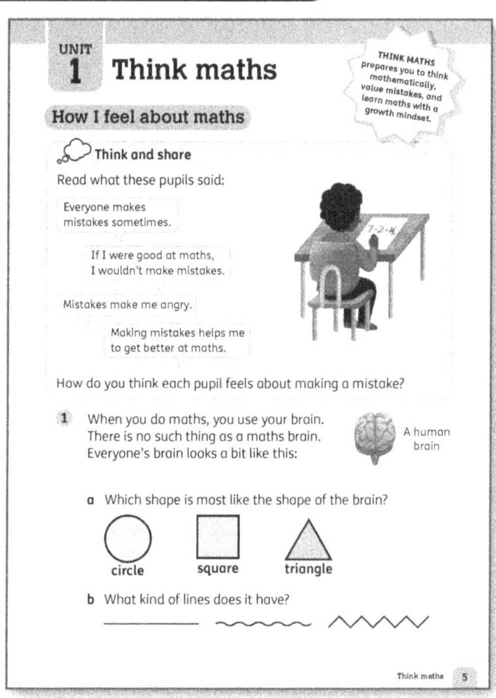

Warm-up

<u>Think and share:</u> Discuss the statements in the speech bubbles on **Pupil Book 2 page 5**. First ask the children to discuss what they think each statement means, and why the person may feel like this. Ask: *Who agrees with this? Who disagrees? Who hates making* mistakes? *Who doesn't mind? Why do we need to make mistakes sometimes?*

Focus

- Ask everyone to imagine they have worked very hard on a maths problem or *question* and then made mistakes or 'got it wrong'. Ask them to think about how that would make them feel.

They can discuss this in pairs first and then share their ideas with the class. If they are shy about offering ideas and suggestions, give them some prompts. For example, list some different feelings on the board: annoyed, happy, excited, curious, frustrated, angry, disappointed, worried.

- Ask the children what they can see in the picture in question 1.
- For part a, you can ask: *Which shape best matches the* brain? *Can you think of any other shape it looks like?* Draw a circle, square and triangle on the board. Ask: *Who thinks the brain looks like this* (pointing to the circle)? *Who thinks it looks like this* (pointing to the square)? *Or do you think it looks like this shape* (pointing to the triangle)?
- For part b, ask the children to describe the brain and the lines they can see.

Mistakes and your brain

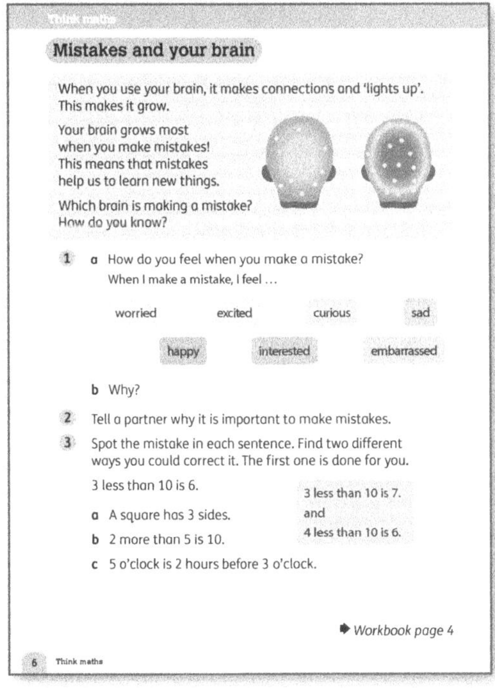

Warm-up

- Read the information at the top of **Pupil Book 2 page 6**. The gaps between neurons in the brain are called *synapses*. Signals pass through these and connections are made: the brain 'lights up' or 'fires'. Ask the children which brain is 'firing'. Emphasise that mistakes cause the most sparks in our brains, and lead to the most growth. Explain that we know the lit-up, colourful brain is making a mistake because we have learnt that the brain 'fires' and makes more connections when mistakes happen.
- Tell the children that when we make a mistake in maths, this is an opportunity to learn. We should never be scared of making mistakes! Getting things right feels good too, but getting things right tells us what we have already learnt. Making mistakes tells us what we are learning *now*. When we make a mistake, we can ask: How did we get this answer? Why doesn't it work? Which part could work? How can we do it in a different way?

Focus

- For question 1, discuss how we should feel about mistakes. Encourage children to understand that mistakes are welcome in our class, because they show that our brains are growing.
- Invite the children to work in pairs on question 2. Emphasise that we celebrate mistakes, because mistakes help us to learn.
- Let the children work in pairs or small groups on question 3, providing support as needed if they cannot yet read. Encourage them to discuss their ideas and reasoning with each other as they work together to find the mistakes. Ask: *What did you learn from looking at these mistakes? How did it help you?*

Follow-up

Workbook 2 page 4 consolidates the ideas introduced and discussed in this unit. The children may need help writing their sentences in question 2.

Interesting mistakes

Many of us carry implicit beliefs and stereotypes about maths. These assumptions may come from our families or friends, from our earlier experiences with the subject, from other teachers, and even from popular culture and the media. Here are a few myths and stereotypes:

- 'Maths is a difficult subject.'
- 'Only some people are truly good at maths.'
- 'If I get things wrong, it means I'm useless at maths.'
- 'Some people have a maths brain and others are better at creative or language subjects.'
- 'Maths is a boys' subject.'
- 'If I ask a question, everyone will know I don't understand, and they might think I'm stupid.'
- 'Smart kids get everything right.'

None of these myths are true, but they still persist – both among children and among teachers. By focusing on how learning takes place, and the importance of mistakes and questions, we can constantly remind children to question these messages.

Answers for Pupil Book 2 page 6

1 a Possible answer: When I make a mistake, I feel curious and interested – but it is not easy to feel happy about it!

b Possible answer: Making mistakes means that I am challenging my brain and learning. Because finding out where I went wrong can help me understand better.

2 Possible answer: Our brains grow most when we make mistakes! This means that mistakes help us to learn new things. Making mistakes helps us find out what we don't understand yet.

3 a A square has 4 sides. A triangle has 3 sides.

b 2 more than 5 is 7. 5 more than 5 is 10. 2 more than 8 is 10. 2 times 5 is 10.

c 5 o'clock is 2 hours after 3 o'clock. 5 o'clock is 2 hours before 7 o'clock. 1 o'clock is 2 hours before 3 o'clock.

Answers for Workbook 2 page 4

1 Individual's answers.

2 a Possible answer: Everyone can do maths, and everyone makes mistakes sometimes.

b Possible answer: Mistakes are important because we learn from them.

Different ways of working

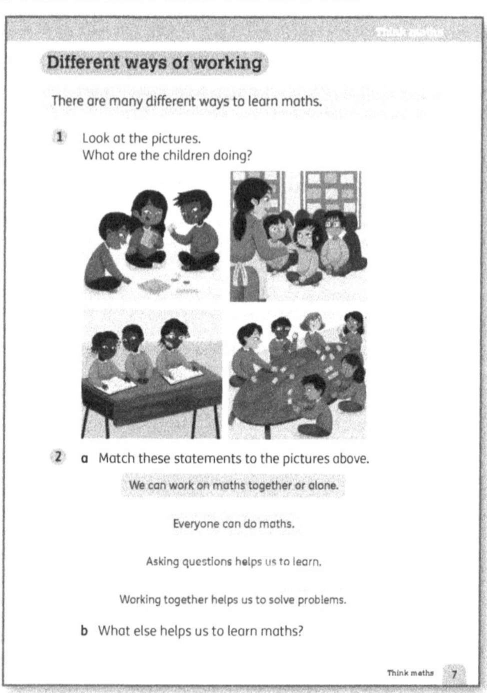

Warm-up

Turn to **Pupil Book 2 page 7**. For question 1, invite the children to say what they see in the pictures. Ask:
- *What are the children doing?*
- *What are they using to help them learn?*
- *Are they working alone or together?*
- *What else do you notice?*

Focus

For question 2, talk about the messages on the strips. Ask the children how each picture could give us one of these messages. For example, in the pictures, children are working together in groups and having fun. Everyone is included. They are listening to the teacher and to each other.

Ask: *What else do these pictures tell us about how people learn maths?* Let the children share any other observations.

Answers for Pupil Book 2 page 7

1 Individual's answers
2 a Possible answers (accept other suggestions with reasons):
- We can work on maths together or alone. – bottom left picture
- Everyone can do maths. – top left or bottom right picture
- Asking questions helps us to learn. – top right picture
- Working together helps us to solve problems. – top left or bottom right picture

b Possible answers: keep trying different way; sharing ideas with a partner or group; not being afraid to try; asking for help; drawing pictures.

End-of-unit check

Discuss with the class what we have learnt about maths in this short unit. Read out these statements and ask everyone to call out whether it is true or false:

- *Maths questions always have only one right answer.* (False – there are usually many ways of working something out, and often more than one possible answer.)
- *If a question is difficult, that means you are bad at maths.* (False – difficult questions mean that you are still learning and working it out; no one is 'bad at maths'.)
- *Some people work slower than others and that is fine.* (True.)
- *Working together helps us to find more ways of working things out.* (True.)

Working with numbers

Learning objectives

- Recognise a number of objects presented in different ways without counting.

- Estimate and count objects and people to 100.

- Recognise a number of objects presented in unfamiliar patterns up to 10, without counting.

- Recognise and use ordinal numbers (first, second, third, and so on).

- Read and write numbers to at least 100 in numerals and words.

- Identify, represent and estimate numbers using different representations, including number lines.

- Count in steps of 2, 3, and 5 from 0, and in tens from any number, forward and backward.

Key words

count   number   tens   ones   number line
estimate   how many   arrange   order
number names (zero to one hundred)
ordinal numbers

Unit introduction

Materials

If using the physical warm-ups: skipping rope, ball

If using the picture warm-up activity: pictures of a bicycle, a three-legged stool, a four-legged table, a starfish and any other objects with features of 2–7 parts that children can subitise (recognise the number without counting)

Teaching guidance

- If appropriate for your class, you can use some of the 'Physical counting warm-ups' (page 25).
- If it is difficult for any children in your class to *count* body parts or complete the physical warm-up activities, we suggest bringing some objects or pictures of objects to the classroom and asking *how many* questions about them. You can use the suggestions above, or any other objects that are available and might be useful. Ask, for example:
  - *How many wheels are there on the bicycle?*
  - *How many legs does this stool have?*
  - *What about this table?*
  - *How many points does this starfish have?*

- As the class settles down, ask them some very simple *number* questions, such as:
  - *How many doors do we have in our classroom?*
  - *How many lights are there?*
  - *How many ears do I have?*
  - *How many fingers am I holding up?*

Invite the children to suggest why we don't always have to count numbers. They may make observations such as: 'Sometimes you just know,' or 'Sometimes you can see.'

Count and show numbers

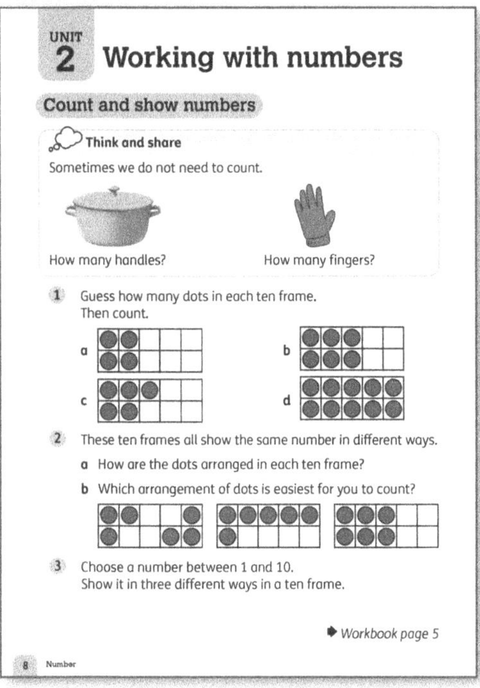

Materials

Photocopies of ten frames (as shown below), with about four ten frames on each sheet; beans or counters; objects such as plastic building blocks or large beads to make groups for children to estimate and count; small containers (such as empty jars or yogurt pots); a tray; base-ten blocks (page 22)

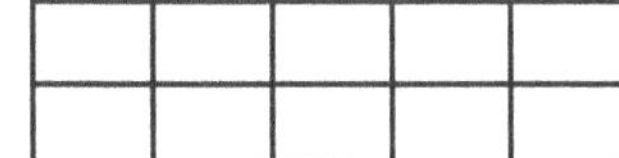

Warm-up

- <u>Think and share:</u> Discuss the two pictures and questions at the top of **Pupil Book 2 page 8**. Children should notice that there are 2 handles on the pot and 5 fingers on the glove. Ask: *How did you know there were 2 handles? Did you have to count? Did you have to count the fingers? Why/why not?*
- Ask the children to count greater numbers. For example, ask: *How many children are in our class today? How many chairs are there in the classroom?*

Focus

- Ask the children to close their eyes. Place 9 counters (or other small objects) on a tray. Cover the tray with a cloth. Tell the children you will remove the cloth just for a moment for them to see what is on the tray. Take off the cloth for a second or two and then cover it again.
- Ask the children to guess the number of objects. Give them a range of options: *Is the number closest to 10, 15 or 20?* Let them debate and discuss their guesses until you have groups for each guess. Then show the objects and let them count. Discuss these questions: *Was it easy or hard to guess? Would it be easier with a greater number or a smaller number?*
- Ask the children to 'guess' how many dots are in each frame in question 1 (you will move on to use the word *estimate* on **Pupil Book 2 page 13**). They then count to check.
- Question 2 shows some different ways to represent the number 6 using ten frames. Hand out photocopies of ten frames and counters. Let the children work independently to *arrange* the number 6 in different ways.
- If any children complete the question easily, give them other numbers to show in different ways. Because the ten frame only has ten squares, there are no other ways to represent 10 than to fill all the squares. However, for numbers between 2 and 8 there are several different possible representations.
- For question 3, children can choose numbers and arrange them in different patterns. Challenge the class to find as many different patterns with their numbers as possible. (You could also do this by drawing or sticking up cut-outs on a blackboard or whiteboard.)

Follow-up

Children work through **Workbook 2 page 5** independently, finding pairs that show the same number and colouring them to match.

Challenge

- Use coins, beans or other objects. Ask the children to choose a number between 10 and 20. They should set out that number of objects and rearrange them in different ways.

- Some children may want to choose larger 2-digit numbers and arrange them in different patterns. They may also start noticing how we can arrange the objects in arrays.

Support

- Work with groups of 0 to 5 objects. Set out a number of objects (1 or 2), and have the child say (or guess) how many there are, then count and check, and then make the same number in another set. Ask: *How many are there if we put one more/one less?* Practise counting together. Make sure each child has a firm grasp of counting from 0 to 10 before moving on to greater numbers. Check understanding of *tens* and *ones*.

- Move on to making 'more tens' using 10-rod base-ten blocks (see page 22). Show 2 tens, 3 tens, 4 tens, and so on. At this point, focus on the numbers 0 to 10, and the multiples of 10 and how we write them.
- When working with the ten frame, ask: *Why would it be difficult to find different ways to arrange 10 counters in a ten frame? Can you do it without the ten frame?*

Ordinal numbers

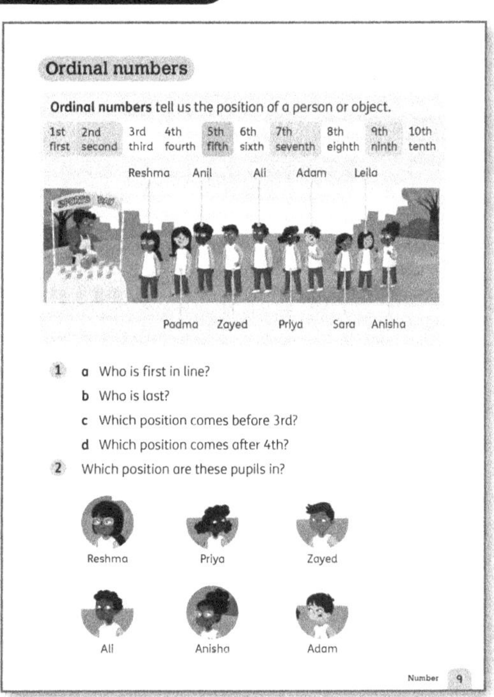

Materials

Large cards with the ordinal numbers 1st, 2nd, 3rd, 4th, 5th, as shown here:

1st	2nd	3rd
4th	5th	

Warm-up

Invite three children to come to the front of the class and stand in a line. Ask: *Who is first in the line? Who is second? Who is third?* Give out the cards to the children. Then move one of the children to a different position. Ask: *Are the cards in the right positions now? Who should have the 1st card? Who should have the 2nd card?* Ask for two more volunteers and add another child to the front and back of the line. Ask them to call out their positions. Let another child come up to be the arranger of the cards.

Focus

- Explain that *ordinal numbers* are numbers that describe an *order*. Ordinals are numbers such as first, second, third, fourth, fifth, and so on.
- Read the information at the top of **Pupil Book 2 page 9**.
- For question 1, invite children to use the ordinal words to describe the children in the picture. Read through the ordinals and explain that they show position. The person in number 1 position is first. The person in the number 2 position is second, and so on.
- As you teach the ordinals in the form 1st, 2nd, 3rd, 4th, and so on, point out that 1st/first, 2nd/second and 3rd/third look different from the others because they have -st, -nd and -rd, whereas all the others have -th. Also help them to notice that for some of the numbers we change the spelling: fifth (not fiveth) and ninth (not nineth).
- For question 2, children identify the positions of the children pictured.

Answers for Pupil Book 2 page 9

1 a Reshma b Anisha   c 2nd   d   5th
2 Reshma: 1st/first   Ali: 5th/fifth
Priya: 6th/sixth   Anisha: 10th/tenth
Zayed: 4th/fourth   Adam: 7th/seventh

Count to 100

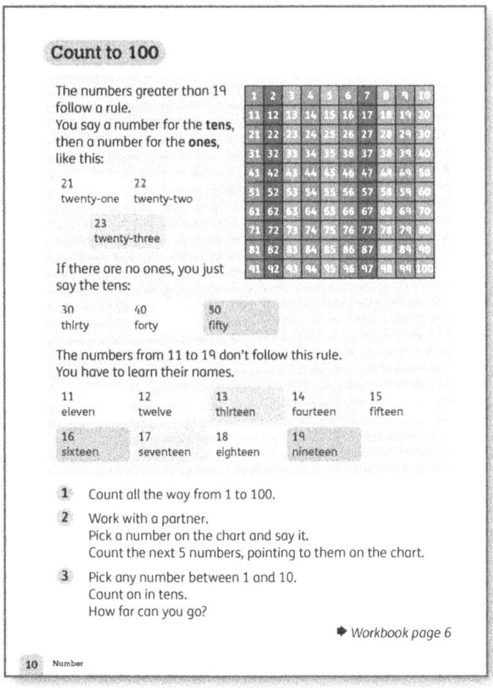

Materials

Place-value cards (see pages 22–23); large 100 chart; number strips; number line marked in tens from 0 to 50; spinners (see ideas below); 1–100 number cards; straws; beans; cups; chalk

Warm-up

There are many suitable activities on pages 25–31. 'Number names and numerals' (page 25) will be most useful. 'Place-value games' (pages 26–27) may be useful.
See also Place value cards (pages 22–23).

Focus

Use some or all of these activities to teach or reinforce counting to 100:

- The children should already be able to count to 100, but they may not know how to read and write numbers beyond 30. Assess what they already know before moving on to this topic. Focus on the pattern of numbers in the twenties and thirties and point out that the same pattern exists in the forties, fifties, sixties, seventies, eighties and nineties. Display number names in words and figures to help the children learn to spell them correctly.
- If needed, you can work up gradually to 100, working first with numbers to 20, then to 50 and then higher. Once the children know the words for the multiples of 10, they should easily learn the pattern of the higher numbers (the multiple of 10, followed by the ones, for example, 42). You can use a number chart or *number line* to go through this for each interval from one multiple of 10 to the next. Point to the numerals, say the names and ask the children to repeat these. It is useful to shade the multiples of 10 on your number chart to highlight these numbers and to allow children to see when they bridge from one ten to another.
- Once you have taught the numbers, point to any number and ask the children to say it. Alternatively, say the name of a number and get different children to point to it on the chart.
- Cover up some numbers on the large 100 chart and ask the children to count, including that number in their count. You can also use this as a quiz to ask them to work out and say the numbers you have covered up.
- You can also use the number chart to count in twos, threes and fives from 0, and in tens from any number, forward and backward. These skills are also practised in Unit 5, where the children skip count when looking at patterns and in Unit 11 (Multiply).
- Use a number spinner made from two paper plates as shown. Children take turns to spin the arrows and then say and write the numbers they have made.

- You can also use place-value cards to make numbers (see pages 22–23). Ask the children to use multiples of 10 together with the unit cards. Arrange these in two piles, facing down. The children draw a tens card and a ones card, make a number and say it, or write its name.
- Reinforce the ordering of numbers orally by asking questions, such as: *What is the number after 35?* (36) *What is the number before 49?* (48)
- Write sets of five or more consecutive numbers (for example: 31, 32, 33, 34, 35) in a random order on the board. Ask children to come to the board and draw lines to join the numbers in order. They should say each number as they join it.
- Give the children a spread of number cards (for example, from 20 to 30, or 75 to 95) in mixed-up order and get them to place these in a row in the correct order. Encourage them to do this from memory, but allow them to refer to the displayed number chart or number line as they need to.
- Draw a 1–50 or 1–100 number grid on an area of the playground using chalk. Give the children instructions to stand on a certain number on the grid and to move to the numbers before or after the first one they stand on.
- To begin counting and developing the concept of different amounts, the children can use straws or beads to represent numbers. A ten can be represented by a bundle of ten straws or by placing ten beans in a cup.

Turn to **Pupil Book 2 page 10**. Discuss the 100 chart with the class. Ask the children to describe what they see. Then count to 50 with the class. Once they have counted to 50, ask questions about the chart, for example:

- *What numbers does the chart show?* (1 to 100)
- *How are the numbers arranged?* (in rows and columns)
- *What makes them easy to read?* (The numbers are in order from smallest to greatest, each row shows ten, the columns are in different colours, for example)
- *What pattern do you notice about each row?* (You can see how the tens change when you go up and down a row and how the ones changes when you go left and right along the rows)
- *How many squares are there in each row?* (10)
- *Are they all the same?* (yes) *How can you be sure?* (You can count them or count the first row and check that all the others are the same length.)
- *Count all the numbers in the first row. What is the last number in the first row?* (10)

- *Now count all the numbers in the second row. How many numbers are in the second row?* (10)
- *How many squares are there if we add together the first row and the second row?* (20)
- *What happens if we count the numbers down the last column (10, 20, 30 …) What pattern is that?* (count up in tens)
- *Why do the numbers make this pattern?* (They go up because each time you add ten.)
- *Do you notice any patterns in the other columns?* (The number in the ones place always matches the number in the row.)

Then have one child pick a number. As a class, say the number and count the next 5 numbers. Give individual children opportunities to do the same with other numbers. Extend the question by asking:

- *What do you notice about the colours when you count on 5 numbers from anywhere on the chart?*
- *Now pick any number. Say the number and count the next 10 numbers. Now what do you notice? How does this work?*

For question 1, count all the way from 1 to 100 with the class.

The children can work in pairs to complete question 2 and question 3.

Follow-up
The children can work through **Workbook 2 page 6** independently. Ensure they know what to do before completing the questions.

Support
- Some children may not yet be fluent in their counting. Revise counting to 10, and then to 20. Extend the number range slowly to 30.
- When you work through the numbers to 100, develop the numbers 20 to 99 before you focus on the numbers 11 to 19. Additional-language learners may find these difficult as the English words for these numbers have a different structure from the words for other 2-digit numbers.

Interesting mistakes
Some children may try to name numbers by saying the ones first. So, for example, they may confuse 12 and 21, 76 and 67. Explore with the class how we can remember which digit shows us the tens and which shows us the ones.

Answers for Pupil Book 2 page 10
1 The children count from 1 to 100.
2 and 3 Individual's answer

Number names

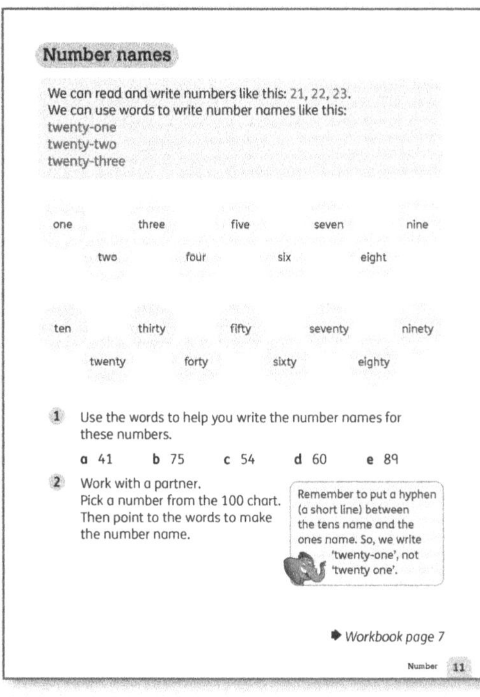

Materials

100 chart; interlocking cubes (page 21) in towers of tens and individual cubes; flashcards with the number names ten, twenty, thirty, up to ninety, and with the numbers zero, one, two, three … up to nine (these could be circular, like the ones on **Pupil Book 2 page 11**, if you wish)

Warm-up

Use the suggestions in 'Number names and numerals' (page 25).

Focus

- Display the 100 chart and a tower of 10 interlocking cubes. Ask: *Which number on the 100 chart matches this number of cubes?* Ask a child to count the cubes and count on the 100 chart. Show one more cube. Ask: *Now which number is it?* (11)
- Say: *I'm going to keep my one cube and add some more tens.* Put two more tens towers together. *How many tens are there now, with my one?* (3 tens) *What is the new number I have made?* (31)

- Continue in the same way with a variety of different numbers, asking questions such as:
  - *What happens if we add more tens to the same number of ones?*
  - *What happens when we change the number of ones?*
  - *What is the greatest number we can make with tens and ones?*
  - *What is the smallest number?*
- Turn to **Pupil Book 2 page 11**. Ask the children to read the number names in the circles. Ask: *Which numbers on the 100 chart can we make from these words? Which numbers can we not make?* (10–19) Point to various numbers on the 100 chart and ask the children to show you which circles you would need to put together to make the *number names*. You can let them practise writing number names on the board, or on their individual boards. Demonstrate the use of the hyphen in the number names.
- The children write the number names for the numbers given in question 1.
- For question 2, the children work in pairs to pick their own numbers from the 100 chart and then identify which circles would make the number names.
- The words in the Pupil Book go up to 99. Ask: *Which number comes right after 99?* (100) *If I have 99 and count 1 more, the next number is … ?* (100) *How do we write this number?* (100) *How do we write the number name?* (one hundred) Demonstrate how to write 100 and *one hundred*.

Follow-up

The children can work through **Workbook 2 page 7** independently. Ensure they know what to do before completing the question.

Challenge

- If any children find writing the number names very easy, give them some riddles to solve. They must write the answer using the number name. Say, for example: *I have 3 tens and 4 ones. What number am I?* (34) *I have 7 tens and 8 ones. What number am I?* (78)
- You can also make this slightly more difficult by introducing addition, for example: *I am the number you get if you put together 2 tens and 3 tens with 4 ones and 3 ones.* (57)

Use a number line

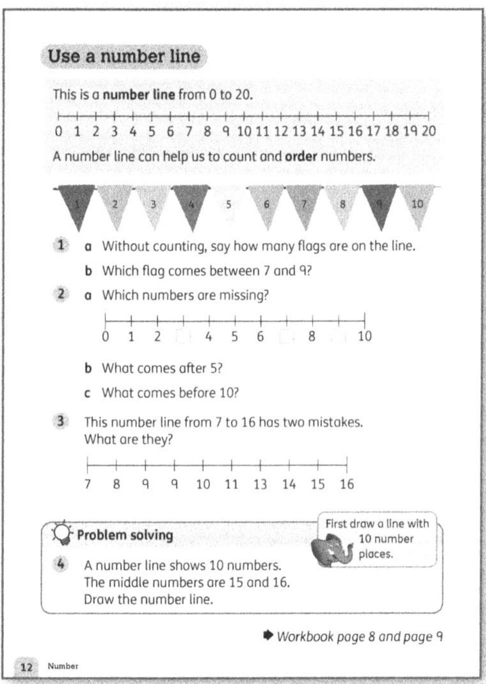

Warm-up
Use 'Missing numbers' (page 26) as a starter for this lesson

Focus
- Turn to **Pupil Book 2 page 12**. For question 1 part a, ask questions about the row of flags:
  - *How many flags are on the line? (10) Did you need to count them? How did you know how many there were?*
  - *Close your eyes. I'm going to call out some numbers and you are going to think about whether you think they are on the number line with the flags. Give me a thumbs up for yes and a thumbs down for no. Could the number 20 be on that number line? (No, thumbs down.) What about the number 2? (Yes, thumbs up.) Would 9 be on that number line? (Yes, thumbs up.) What about 19? (No, thumbs down.)*
  - *Open your eyes. Can someone explain how they knew, without looking, which numbers would be on the number line?*
- For part b, ask various questions using the words *before*, *after* and *between*. For example, ask: *Which number comes between 7 and 9? (8) Which number comes before 2? (1) Which number comes after 2? (3)*
- Next, say: *Imagine that the row of flags continues for another ten numbers. Which number would be the last one on that row? (20) Which numbers could you see on the flags? (11, 12, 13, 14, 15, 16, 17, 18, 19, 20) Which numbers would not appear?*
- Talk about the number line with the missing numbers in question 2. Discuss the questions with the class.
- Discuss the number line in question 3. It has two errors: the number 9 is repeated, and the number 12 is missing. Explain that when you draw a number line, it has to have even spaces between the numbers, and all the numbers must follow the same rule. So, if each number is one more than the one before, we have to continue in the same way with all the numbers.
- Problem solving: Discuss question 4 with the class. Approach it as a number talk (see 'Number talks', pages 17–18).
  - First present the problem, then ask the children to think quietly how they could solve it. They can use hand signals (see page 17) to show when they have an idea.
  - Then give them time to discuss with a partner their suggestions for how to work it out.
  - Finally, call on individual children to share their strategies with the class. Discuss each strategy as they demonstrate for the class. Continue using hand signals so that children can engage with the discussion and take turns to share their individual answers.
- Draw a number line on the board and write the term *number line*. Explain that when we draw a number line, it helps us to see a row of numbers with spaces between them. The numbers must be arranged in order. Point out that even if you are counting backwards, you write out the sequence from left to right.

Follow-up
The children can then work through **Workbook 2 pages 8–9**.

Interesting mistakes
For Pupil Book question 3, it is likely that some children will say that it is a mistake that the number line starts on a 7. This is a good opportunity to explore different kinds of number line. We can make a number line show any range of numbers. We can also use it for skip counting, which you can demonstrate with a number line of tens or fives.

Answers for Pupil Book 2 page 12
1 a 10   b 8
2 a 3, 7, 9   b 6   c 9
3 9 appears twice. 12 is missing.
4 Problem solving:

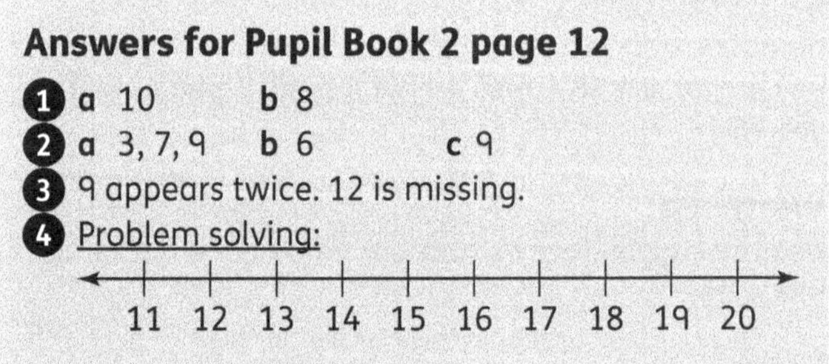

Answers for Workbook 2 page 8
1 a 27, **28**, **29**, 30   b 32, **31**, 30, **29**
   c 37, **38**, **39**, 40   d 43, **44**, **45**, 46
   e 47, **46**, 45, **44**   f **29**, 28, **27**, 26
   g 50, **49**, **48**, 47   h 39, 38, **37**, **36**
   i 14, 16, **18**, 20   j 21, **23**, 25, 27
2 37

1 The children should choose numbers in the correct ranges and mark them correctly on the number lines.

Estimate and count

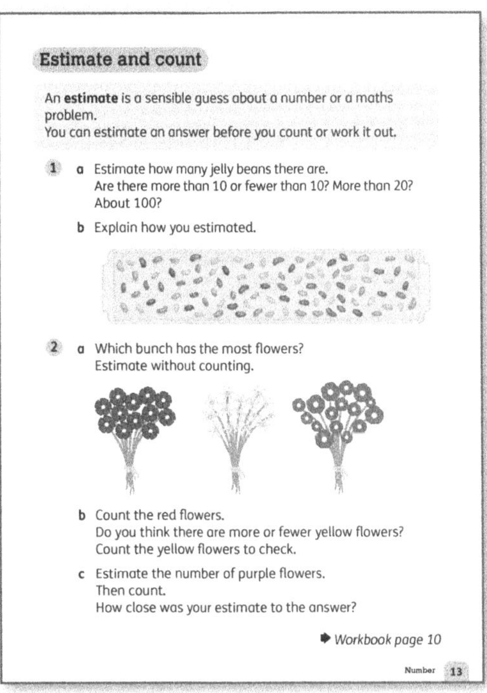

Materials

Bottle (that can be sealed); two types of object to count that will fit inside the bottle, such as counters and beans, or marbles and grains of rice; blank grid with 50 squares, as shown here:

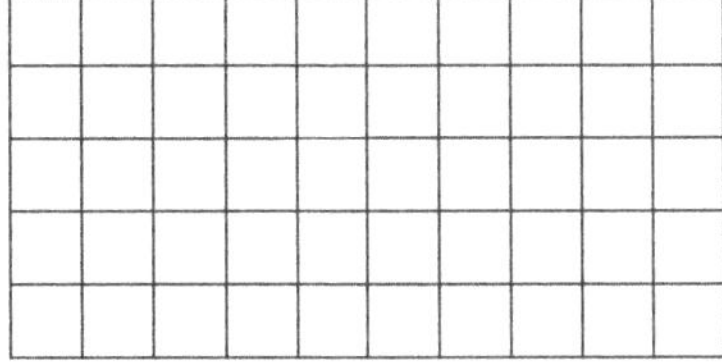

Warm-up

Play a game in which up to four children display both hands (to indicate 10), while you show a number less than 10 using your own fingers. Let the children take turns to count in tens and ones and then say the number shown.

Focus

- Place a number of beans or counters in the bottle and seal it. Pass the bottle around the group and ask the children to guess how many are inside. Empty the bottle and count to see who is the closest. Repeat this activity a number of times. Start with a small number and increase the number of beans. Do not exceed 50. Repeat the activity using larger or smaller objects, such as marbles and rice. Observe whether the children change their estimates accordingly.

- Use a blank 50 grid and a counter. Make sure all the children can see clearly. Place a counter in any position on the grid and ask the children to guess what the number is. Count in a range of ways to see which estimates were close. Ask the children who were correct to explain their strategy.

- Arrange children in groups facing each other. They should have a big pile of counters between them. The children take turns to take a handful of counters. The one who is taking a handful holds out their hand for a short time and then closes it. Each child estimates how many. They then count and discuss which guesses were most accurate.

- Turn to **Pupil Book 2 page 13**. Remind the children what we mean by the word *estimate*. An estimate is a sensible guess based on what we already know.

- For question 1 part a, ask the children how many jelly beans they think there are in the picture. Say: *Are there more than 10 or less than 10? Are there more than 20? Do you think there are about 100?*

- For question 1 part b, the children discuss how they estimated.

- The children answer the questions in question 2 by estimating and counting.

Follow-up

Ensure the children understand the instructions before they complete the activity on **Workbook 2 page 10**.

Interesting mistakes

Children may express concern that they cannot guess a number correctly and exactly, and that it's impossible to get the 'right' answer. Some children may be convinced they should try to count the groups of objects as fast as possible. Invite different children to suggest their own strategies for estimating (for example, thinking about how many tens it looks like, or how much space there is between objects). Explain that when you estimate, the point isn't to find the correct answer but rather to direct us towards a likely or possible answer.

Answers for Pupil Book 2 page 13

1 a about 100   b Individual's answers

2 a Individual's estimates

b Count: 12 red flowers, Individual's estimates, Count: 15 yellow flowers (more)

c Individual's estimates. Count: 16 purple flowers

Answers for Workbook 2 page 10

1

a	Estimate: 20	Count: 20
b	Estimate: 20	Count: 27
c	Estimate: 50	Count: 40
d	Estimate: 20 or 50	Count: 33
e	Estimate: 20	Count: 21
f	Estimate: 50	Count: 41
g	Estimate: 10	Count: 13
h	Estimate: 10	Count: 14

Carry out these activities to assess the children's counting:

- Ask the children to count from a given number on the 100 chart. (For example, start at 39 and count 40, 41, 42, …)
- Cover any set of numbers. Ask the children to count from a number before the covered numbers to a number after. (For example, cover 62, 63 and 64, the children count from 61 to 65)
- Use the 100 chart on **Pupil Book 2 page 10**. Call out a number. The children cover that number on their chart with a counter. Repeat with some more numbers. After ten numbers are covered, check that they have covered the correct numbers.

Use the following estimating activities on pages 28–29 to consolidate and assess number work from this unit:

- 'Estimate with number lines' (page 29). Ask: *What number is this? How did you decide?* (For example, have an arrow on a sticky note pointing to 63 on a blank number line marked from 0 to 100. A child might say 'I think this is about 60 because it is more than half way between 0 and 100 but close to this line' where they demonstrate pointing to the 60 mark.)
- 'Estimate' (page 28). Ask the children to estimate the items in a jar (20 to 50 items) while showing a smaller jar with 10 items in it. Ask: *How did you decide?* (The children explain how they used the items in the smaller jar to estimate how many items were in the larger jar using proportional reasoning.)
- 'Estimate with dot cards' (page 29). Show the children prepared dot cards and ask them to estimate the dots. Ask: *How did you work this out?* (The children explain that they used the arrays/ groups the dots are arranged in to find the number of dots. Encourage the children not to count.)
- 'Estimate using 100 squares' (page 29). Display 100 squares with different patterns of shaded squares and ask the children to estimate (to the nearest 10) how many squares are shaded or unshaded. Ask: *What strategies did you use to estimate?* (The children explain that they used the groups they could see in the patterns to estimate the total. For example, for a pattern showing 5 rows of 10 shaded squares they could say 'Each row has 10 squares shaded and there are 5 rows, so there are $5 \times 10 = 50$ shaded squares.')

UNIT 3 Place value

Learning objectives

- Recognise the place value of each digit in a 2-digit number (tens, ones).
- Understand and explain the value of each digit in a 2-digit number.
- Use place value and number facts to solve problems.
- Compare and order numbers from 0 to 100.
- Use <, > and = signs.
- Compose, decompose and regroup 2-digit numbers using tens and ones.

Key words

place value   place-value table   tens   ones
digit   even   odd   compose   decompose
< sign   > sign

Materials
Photocopies of ten frames (page 21); base-ten blocks (page 22); counters; whiteboard and markers

Teaching guidance
Use a number talk to open this unit (see 'Number talks', pages 17–18). Remember the steps in presenting a number talk:

- **Step 1: Present the problem and give time for independent thinking**, with hand signals to indicate when the children have ideas to share. Avoid taking the first ready answer – give plenty of 'think' time, until many children are giving the hand signal that they have one or more ideas to share.
- **Step 2: Sharing in pairs**, with emphasis on how they solved it. The children use hand signals to indicate when they are finished.
- **Step 3: Individual children share their strategies**, by modelling their way of solving the problem for

the class. The children use hand signals to indicate whether they agree or disagree.

- **Step 4: Discussion**. Here, other children get an opportunity to share their way of doing it, and children discuss and compare their strategies.

Here is a problem to use for this number talk:

- Write a number on the board and ask the children to think of as many different ways as they can to represent that number. You may need to explain that *represent* means show; they can think of any way to show what the number means, or how to make the number. You might want to start with the number 24.
- Follow the talk steps. When you get to the step of having the children share their ideas with the class, write each child's idea on the board so that you create a visual poster showing a variety of ways to show a number. If you notice that children are all gravitating towards the same strategy (for example, a range of addition or subtraction calculations) encourage them to find totally different ways to show the number. For example, say: *What about using some of these* (showing counters or ten frames)? *What about in a table? Did anyone show it using tens and ones?* (Check the children's understanding of *tens* and *ones*.)
- Here is an example of what your board might look like:

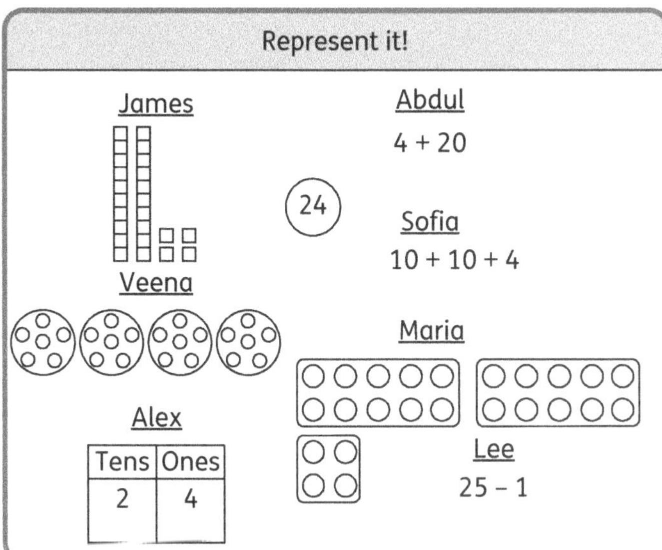

- You can repeat this with other numbers.

Number talks are a 'low floor, high ceiling' activity. This means that all the children in your class should be able to engage with the activity at their own level. You may have children who are already thinking in terms of multiples and factors (24 is double 12, or it is 3 eights, or it is half of 48), whereas others will still be trying to work out how to partition the number. Allow all the children who answer to explain their thinking. If someone's reasoning is incorrect, allow others to explain why they agree or disagree.

Tens and ones

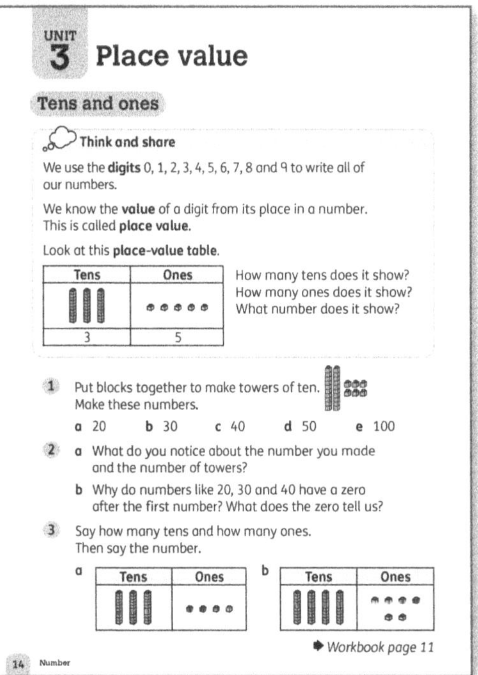

Materials

Place-value tables (page 23); place-value cards (page 22); interlocking cubes or base-ten blocks (pages 21 and 22); 1–100 number cards

Warm-up

Remind children of the activity you did in the unit introduction: *We looked at the number 24 and we showed it in many different ways. Let's look again at the number.* Write it on the board: 24. Point to the first *digit* (2). Ask: *What does the 2 show? What does the 4 show? Let's look back at the different ways we showed 24. Which of our representations showed the tens and the ones?* (In the example given on the left, James, Abdul, Sofia, Alex and Maria have all represented 24 as tens and ones.) Point out to the children that way we write numbers from 10 to 99 always shows the tens first, followed by the ones.

Focus

- Use a *place-value table*. Use base-ten blocks to represent the number 31, with 3 10-rods and 1 ones cube. Ask the children to identify the number you have made. Ask them to explain how they worked it out. Have a child come up and write the number on the board. Repeat with other numbers.
- Use 'Place-value games' (page 26) and 'Show it four ways' (page 27).
- <u>Think and share</u>: Read through the information and look at the place-value table on **Pupil Book 2 page 14** with the children.
- For question 1, provide the children with interlocking cubes to make tens. As a class, make the multiples of 10 listed in the question.
- For question 2 part a, ask the children what connection they notice between the number and how many sticks (or towers) of ten they used to show it? Ask: *What is the clue in the number that tells us how many tens it is?*

- Part b raises the question of the position of zero. This is a good opportunity to look at the way zero functions in different numbers. Use place-value cards to compose the number 23 (using a 20 and a 3). Ask:
  - *What does the 2 show us? (2 tens) What does the 3 show us? (3 ones)*
  - *Let's take away 1 one from 23. What number is that?*
  - *What is the number that is 1 less than 23? (22)*
  - *Now we have 2 tens and 2 ones. What if we take away another one? (21 – 2 tens and 1 one)*
  - *Now let's take away another one. Now we have 2 tens and how many ones? (None).*
- Make an intentional error, taking away the ones but writing 20 as as 2. Ask: *Is this correct? Can I just use a 2 to show 2 tens?* Pretend that you really don't understand and let the children correct your mistake.
- Stimulate a discussion about why we have to represent twenty as 20, not as 2. Guide the children to understand that we have to use zero as a placeholder so that the other numbers keep their correct *place value*. Reinforce by building more 2-digit numbers, including multiples of 10, with place-value cards.
- The children identify the tens and ones in each place-value table in question 3.

Follow-up

Work through one or more examples on **Workbook 2 page 11** with the class. Some children may be able to complete the page independently, while others may need assistance.

Challenge

Give the class some 'riddles' and let them work out what the number is. Check that they understand what is meant by *even* and *odd* numbers and include these terms. Say, for example:

- *I am between 20 and 30, I am even and I am greater than 25. What number could I be?* (26 or 28)
- *I am a number with 3 in the ones place and I am between 40 and 50. What am I?* (43)
- *I have two digits, 6 and 3. I am between 30 and 40. What number am I?* (36)

Support

For children who need more practice, continue making more numbers using place-value tables, drawing attention to the number of tens and ones. The 'Show it four ways' sheets (page 27) are particularly helpful for this. Give practice with numbers under 20 and, once they gain more confidence, move on to numbers up to 50.

Answers for Pupil Book 2 page 14

<u>Think and share:</u> three tens; five ones. 35, thirty-five

1 a two 10-rods   **b** three 10-rods
   c four 10-rods   **d** five 10-rods
   e ten 10-rods

2 a The number of 10-rods is the same as the tens digit in the number.
   b There are no zeros in the number. The number is a number of tens, with no ones.

3 a 3 tens 4 ones, 34 (thirty-four)
   b 4 tens 6 ones, 46 (forty-six)

Answers for Workbook 2 page 11

1 a 48   **b** 3 beads on T and 3 beads on O
   c 49   **d** 2 beads on T and 8 beads on O
   e 50   **f** 3 beads on T and 5 beads on O

2 a 4 tens 0 ones [Provided as an example]
   b 5 tens 9 ones
   c 7 tens 3 ones
   d 9 tens 0 ones

3 a 4   **b** 5

Make and break numbers

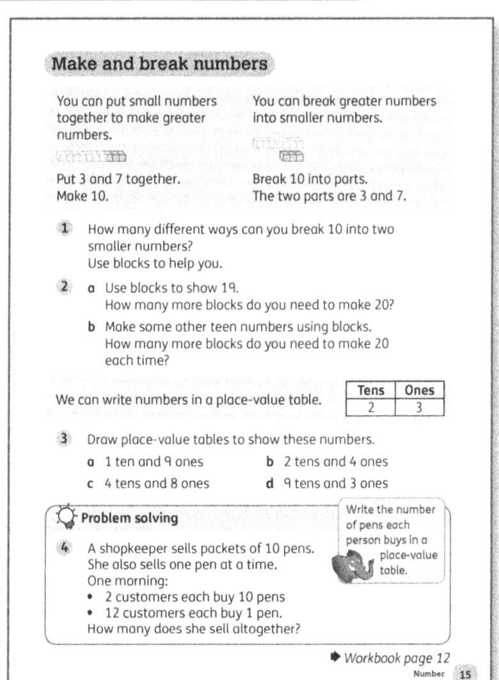

Materials

Interlocking cubes (page 21) in different colours, or counters of two different colours and ten frames (page 21)

If you have any colour-blind children in your class, it is useful to use black and white counters in ten frames rather than coloured counters. Alternatively, use black and white cubes.

Warm-up

Turn to **Pupil Book 2 page 15**. For question 1, build a stick of 10, composed as shown at the top of the page – 3 of one colour and 7 of another, put together to make 10 (or use ten frames and counters). Ask:

- *What numbers can you see here?* (3, 7 and 10)
- *What smaller numbers have I used to make a greater number?* (3 and 7 to make 10)
- *What other way could I show the same number?*

Give the class some time to show different ways to make a number less than 10. They might choose to count out ones or to count out a stick of 10 and take away 1.

Focus

- For question 2 part a, ask the class how to show the number 19. Let one of the children compose it using more of the interlocking cubes added to your 10. Ask: *How many more cubes do you need to make 20?* (1) Let them show you. Then ask: *Did you have to count? Is there another way to work it out?*
- For part b, invite the children to make some other teen numbers (14, 13, 18, and so on) using the interlocking cubes. For each number they make, ask how many more cubes they need to make 20. Ask them to explain how we do this.
- Demonstrate how to draw a place-value table (see page 23) before the children draw their own tables for question 3.
- Problem solving: Before you ask children to work on question 4, present the information alone. Explain that the shopkeeper sells pens like this: in packs of 10, and as single pens. Draw a picture on the board to show a packet of 10 pens and a single pen. Ask:
  - *How many do you get if you buy a packet of 10?* (10)
  - *How many do you get if you buy 2 packets of 10?* (20) *How many are in 3 packets of 10?* (30)
  - *How many packets should you buy if you need 50 pens?* (5)
  - *What if you need this many pens?* (Write the number 18.) (1 packet and 8 pens)
- Then present the question as a numberless question (see 'Numberless word problems' on page 18). *Some customers buy packets. Some customers buy 1 pen each.*
  - *What question could we ask?*
  - *What information would we need to work it out?*
- Work through the steps of discussion in talking about the numberless question before you give the rest of the information from question 4.

Follow-up

Work through some examples on **Workbook 2 page 12** before asking the children to complete the page independently.

Challenge

Ask: *Instead of breaking 10 into two smaller numbers, how many different ways can you break it into three smaller numbers?* Try the same exercise with numbers between 10 and 20.

Answers for Pupil Book 2 page 15

1 1 and 9, 2 and 8, 3 and 7, 4 and 6, 5 and 5

2 a 1 more b Any answers from: 13, 7 more; 14, 6 more; 15, 5 more; 16, 4 more; 17, 3 more; 18, 2 more

3

a

Tens	Ones
1	9

b

Tens	Ones
2	4

c

Tens	Ones
4	8

d

Tens	Ones
9	3

4 Problem solving: 32

Answers for Workbook 2 page 12

1 a 7 tens 3 ones, seventy-three, 73
   b 3 tens 6 ones, thirty-six, 36
   c 4 tens 5 ones, forty-five, 45
   d 5 tens 6 ones, fifty-six, 56
   e 8 tens 2 ones, eighty-two, 82
   f 6 tens 8 ones, sixty-eight, 68

Compose and decompose

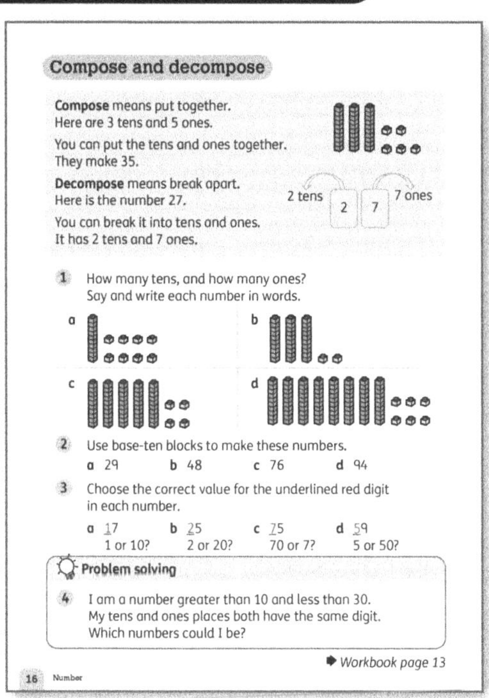

Materials

Wooden ice cream sticks or strips of card to make bundles of tens and ones; rubber bands; interlocking cubes (page 21); place-value cards (pages 22–23); beans; cups

Warm-up

The children work in pairs for these activities:

- Give each pair 20 sticks or strips of card and a rubber band. They group 10 items together with the rubber band to make a bundle of 10. One child says a number between 10 and 20 and the other child uses the bundle of 10, together with other 'ones', to represent the number.

  The children then carry out the reverse process: one child uses the materials to represent a number and the other child counts the number of objects to find the number.

  Repeat this using place-value cards to relate what you are doing to tens and ones and place value.

- Give each pair 20 beans and two cups. They put 10 beans together in a cup to make a group of 10. One child says a number between 10 and 20 and the other child uses the group of 10, together with other beans, to represent the number.

  The children then carry out the reverse process: one child uses beans to represent a number and the other child counts the number of beans to find the number.

  Repeat this using place-value cards to relate what you are doing to tens and ones and place value.

- Extend the two activities. Children use the same materials to represent numbers between I and 50.

Focus

- Write the word *compose* on the board. Ask whether the children know this word. Explain that a composer is someone who makes music: they put together all the notes to make the song. *Composing* means putting things together. *Decomposing* means breaking things apart. Demonstrate how to compose and decompose the number 35 using interlocking cubes as shown on **Pupil Book 2 page 16**.
- The children inspect each number in question 1 and say how many tens and how many ones there are. They can compose the numbers using interlocking cubes. They say the number names and write them in words.
- The children use interlocking cubes to make the numbers in question 2.
- For question 3, talk about what the red digit shows in each number. (The digit is also underlined, in case you have colour-blind children in your class.)
- Problem solving: Discuss the riddle in question 4 with the class. Use the same format as you use for your number talks, giving individual think time, sharing time and discussion time as a class.

Follow-up

Ensure the children are confident with using place-value cards before they complete **Workbook 2 page 13**.

Challenge

You can give additional problem-solving questions for more challenge:

- *I am a number greater than 20 and less than 35. I have a 0 in my ones place. What number am I?* (30)
- *I am a 2-digit number greater than 95. My ones digit is 2 less than my tens digit. What number am I?* (97)

Interesting mistakes

A common mistake is to mix up the tens and the ones, for example to confuse 81 and 18, or 45 and 54. Explain that this is a very interesting mistake, because it shows how we use place value.

Draw an unlabelled place-value table on the board. Say: *I am thinking of a number. It has some tens and some ones. One of the digits is a 5 and one of the digits is a 3. What numbers could it be?* Children should identify 53 and 35. Ask:

- *If it is 35, how many tens does it have and how many ones?* (3 tens and 5 ones)
- *If it is 53, how many tens does it have and how many ones?* (5 tens and 3 ones)
- *Which column does the tens go in?* (the left one)

Remind the children that the tens always go in the place before the ones.

Answers for Pupil Book 2 page 16

1 a 1 ten 8 ones, eighteen
    b 3 tens 2 ones, thirty-two
    c 5 tens 4 ones, fifty-four
    d 8 tens 6 ones, eighty-six

2 a 2 tens 9 ones   b 4 tens 8 ones
    c 7 tens 6 ones   d 9 tens 4 ones

3 a 10   b 20   c 70   d 50

4 <u>Problem solving</u>: 11 or 22

Answers for Workbook 2 page 13

1 a 30 + 7 = 37 [Provided as an example]
    b 40 + 5 = 45
    c 80 + 3 = 83
    d 1 + 70 = 71

2 a seventy-two   b thirty-eight
    c eighty-eight   d eighty
    e sixty-nine   f forty-nine

	Tens	Ones
a	7	2
b	3	8
c	8	8
d	8	0
e	6	9
f	4	9

Use the = sign

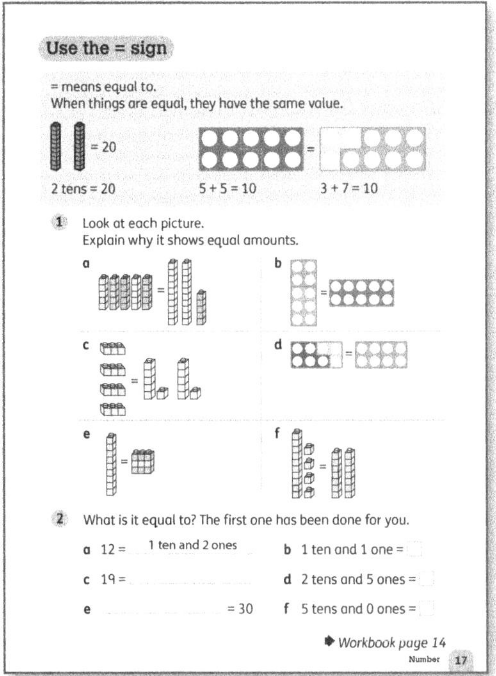

Materials
Numicon (page 21–22) or interlocking cubes (page 21)

Warm-up
- Ask the children to suggest a number between 30 and 40, such as 37.
  Write the number on the board and then write an equals sign and a large box.
  $37 = \square$
  Ask: *What is equal to 37?*
- Approach this as a number talk, giving the children time to come up with their own ideas (see 'Number talks', pages 17–18). Follow a similar procedure to the one used in the unit introduction (see page 42). Make sure that each time you write one of the ideas on the board, you use the equals sign. Your board may look something like this:

<table>
<tr><td colspan="2" align="center">What is = to 37?
Equal</td></tr>
<tr><td align="center">Linda
37 = 30 + 7</td><td align="center">Aaliyah
37 = 35 + 2</td></tr>
<tr><td align="center">Nicholas
37 = 40 − 3</td><td align="center">Mia
37 = 3 tens and 7 ones</td></tr>
<tr><td colspan="2" align="center">Noor
37 = 38 − 1</td></tr>
</table>

Focus
- Ask: *What symbol have we used to show equals?* Point out the equals sign. Ask: *Who can tell me what this sign means?* Let the children share their ideas.

Lead them to understand that = means equal to. It tells us about the relationship between what is on each side of the sign.
- The children look at each pair of pictures in question 1 on **Pupil Book 2 page 17** and explain why they show equal amounts. Discuss what they see on one side of each picture and compare it to the other side.
- For question 2, the children complete number sentences so that each side of the = sign shows the same number: expressed as a number on one side and decomposed into the tens and ones on the other side.

Follow-up
Discuss each example on **Workbook 2 page 14** with the class before letting them work independently to complete the questions.

Support
Prepare sets of Numicon or interlocking cubes and ask the children to work out whether they are equal.

Interesting mistakes
The plus sign and the minus sign are operation signs (also called *operators*). This means they tell us to perform an operation. The plus sign tells us to add and the minus sign tells us to take away. All too often, this means that children see one of these frameworks:

$$\_\_ + \_\_ = \_\_, \text{ or } \_\_ - \_\_ = \_\_$$

This can lead children to believe that the equals sign is also a kind of operation sign, telling them to 'find the answer'. However, in this lesson, we are showing children that the equals sign is not an operation sign. It is a sign that tells us the relationship between two expressions. It is a sign that tells us two things have the same value.

Answers for Pupil Book 2 page 17
1. a $5 + 5 + 5 + 5 + 5 = 25$, $10 + 10 + 5 = 25$
   b $4 + 6 = 10$
   c $3 + 3 + 3 + 3 = 12$, $5 + 1 + 5 + 1 = 12$
   d $5 + 3 = 8$, $4 + 4 = 8$
   e $6 + 3 = 9$, $3 + 3 + 3 = 9$
   f $10 + 1 + 1 + 1 + 1 = 14$, $7 + 7 = 14$
2. a 1 ten and 2 ones [Provided as an example]
   b 11   c 1 ten and 9 ones
   d 25   e 3 tens 0 ones
   f 50

Answers for Workbook 2 page 14
1. a 3 tens   b column of 5 circles   c 8
   d 1 ten 2 ones
2. Individual's answers

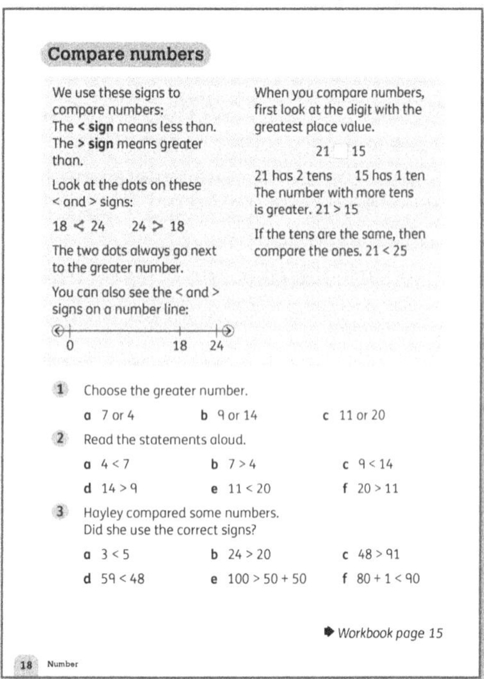

Materials

Base-ten blocks (page 22) or ten frames (page 21)

Warm-up

- Show two numbers using base-ten blocks. For example:

Use the number talk approach (see pages 17–18) to work through the following problem:

Two children each have some blocks. Malia has this many and Joseph has this many. Who has more?

- Give some 'think time' and wait until all the children have some ideas about how to work out which number is greater. Let them talk to each other and then, finally, let them share their ideas with the class. Ask:
  ○ *Who has more?* (Joseph)
  ○ *How do you know?* (because I counted them, because I can see more, for example)
  ○ *Did you have to count every block?* (No. You can count up in tens when you count the 10-rods and then on in ones for the ones cubes.)
  ○ *How did you work out the number of blocks each child has?* (I counted up in tens and then on in ones.)

Focus

- Draw the > *sign* and < *sign* on the board and ask whether anyone knows how we use them.
- Show the < and > signs on **Pupil Book 2 page 18**. Explain that we call these signs *greater than* and *less than*. The open part of the sign always opens towards the greater number. Ask: *Which number has more blocks – Joseph's or Malia's?*
- Demonstrate how to draw the signs. Work through the example at the top of the page. Show how we always compare the tens first.
- For question 1, children choose the greater number. Also ask: *Is there another way we can say this?* Encourage them to see that if 7 is greater than 4, 4 is less than 7.
- Children read out the statements in question 2 using the signs.
- Ask children to look at each statement in question 3 and discuss with the class whether Hayley has used the correct sign or not.

Follow-up

Discuss each example on **Workbook 2 page 15** with the class before letting them work independently to complete the questions. Assist as needed.

Support

You can give children these example questions to provide extra practice.

1 Copy. Fill in the < or > sign.

   a 3 ___ 9   b 8 ___ 4   c 11 ___ 12

   d 24 ___ 28

2 Use the < or > sign. Copy and complete these number sentences.

   a 6 ___ 10   b 4 ___ 12   c 16 ___ 7

   d 19 ___ 5   e 20 ___ 30   f 50 ___ 40

   g 30 ___ 100   h 100 ___ 90

Answers for Pupil Book 2 page 18

1 a 7   b 14   c 20
2 a 4 is less than 7   b 7 is greater than 4
   c 9 is less than 14   d 14 is greater than 9
   e 11 is less than 20   f 20 is greater than 11
3 a yes   b yes   c no
   d no   e no   f yes

Answers for Workbook 2 page 15

1 a <   b 10 ones (or 1 ten) > 6 ones
2 a <   b <   c <
   d >   e >   f <
3 a =   b <   c <   d <
4 a ✓   b ✗   c ✓
   d ✓   e ✗   f ✓
5 a 55   b 31   c 49   d 95

Order numbers

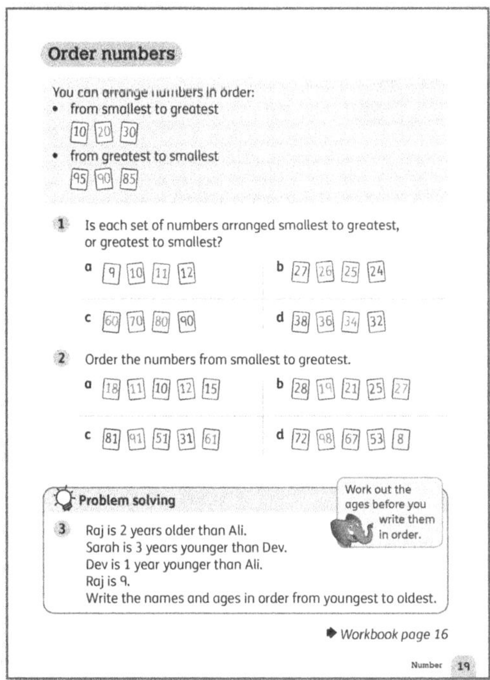

Materials

Number cards with 2-digit numbers; place-value cards (page 22)

Warm-up

Ask questions such as: *What is the number after 65?* (66) *What is the number before 89?* (88) Ask the children to build the answers using place-value cards.

Focus

- Show a mixed set of number cards and ask the children to work out how to put them in order. Initially, use sets of consecutive numbers (for example, 23, 25, 24, 26). Then move on to mixed sets where the numbers are not necessarily consecutive.
- Once the children have had some practice doing this with sets of number cards, work through **Pupil Book 2 page 19** with the class.
- The children look at each set of numbers in question 1 and say whether it is arranged smallest to greatest (ascending) or greatest to smallest (descending).
- They rewrite each set of numbers in question 2 in order from smallest to greatest.
- Problem solving: In question 3 the children can work in pairs. Some may find it helpful to write down the names of the children and then write the ages as they work through the clues of the puzzle. Encourage the children to work out the ages before comparing them.

Follow-up

For **Workbook 2 page 16**, discuss each example with the class before letting them work independently to complete the questions. Assist as needed.

Answers for Pupil Book 2 page 19

1 a smallest to greatest   b greatest to smallest
c smallest to greatest   d greatest to smallest
2 a 10, 11, 12, 15, 18   b 19, 21, 25, 27, 28
c 31, 51, 61, 81, 91   d 8, 53, 67, 72, 98
3 Problem solving: Sarah 3, Dev 6, Ali 7, Raj 9

Answers for Workbook 2 page 16

1 a 16, 19, 23, 28, 32   b 15, 23, 28, 31, 36
c 23, 29, 32, 34, 43   d 14, 32, 33, 41, 43
e 37, 38, 45, 47, 49
2 a 32, 23, 21, 19, 14   b 50, 46, 45, 34, 32
c 38, 23, 21, 19, 14   d 50, 48, 24, 23, 12
e 45, 41, 37, 35, 24

End-of-unit check

Carry out these activities to assess the children's understanding of place value:

- Ask the children to show you given numbers using Numicon or other concrete materials. Also give them some numbers represented using materials and ask them to say or write the number.
- Prepare a range of statements about numbers and read these out to the class. The children must say whether the statement is true or false. If it is false, you may like to ask them to correct it. You could use statements such as these:
  - *43 is 10 more than 33.* (true)
  - *72 is between 60 and 70.* (false)
  - *45 is 1 less than 44.* (false)
- Show the children a 2-digit number and ask them to decompose it into tens and ones.
- Write sets of five or so consecutive numbers (for example, 81, 82, 83, 84, 85) in a random order on the board. Ask the children to come up and to draw lines to join the numbers in order. They should say each number as they join it.
- Give the children a spread of number cards (for example, from 50 to 60, or 55 to 65) in a mixed-up order and get them to place these in a row in the correct order. Encourage them to do this from memory, but allow them to refer to the displayed number chart or number line if needed.

2D and 3D shapes

Learning objectives

- Identify and describe the properties of 2D shapes, including the number of sides.

- Identify, describe, sort and name 3D shapes by their properties, including the number of edges, vertices and faces.

- Identify 2D shapes on the surface of 3D shapes (for example, a circle on a cylinder and a triangle on a pyramid).

- Compare and sort common 2D and 3D shapes and everyday objects.

- Understand that a circle has a centre and any point on the boundary is at the same distance from the centre.

Key words

2D shape   3D shape   sphere   cube   cuboid
pyramid   cone   prism   cylinder   edge   face
polygon/regular polygon   properties   side
vertex   vertices   circle   square   triangle
rectangle   corner   boundary   centre   length

Unit introduction

Materials
A range of 2D and 3D shapes; items or pictures that show real-life examples of the 2D and 3D shapes you are working with

Teaching guidance
Use one or more of the 'Identify 2D and 3D shapes' activities (page 31).

2D shapes

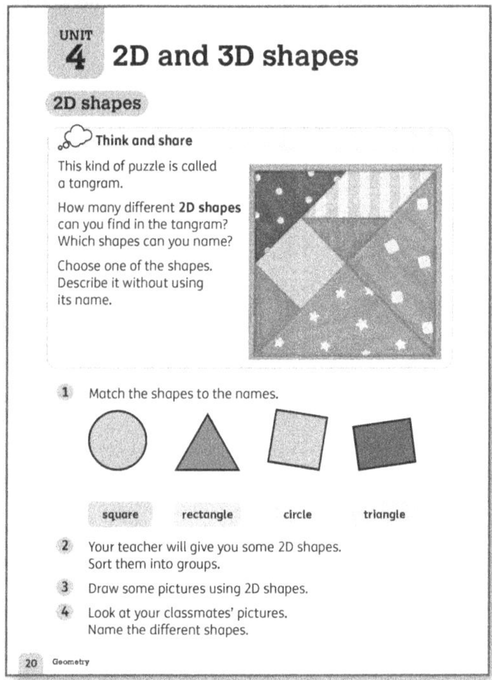

Materials
A set of tangram shapes; 2D shapes for the children to sort (rectangles, triangles, circles and squares)

You may have a tangram puzzle made of wood or plastic. If not, copy the 'Shape jigsaw' on page 24, or do a internet search for a tangram creator to find an online version.

The tangram puzzle is made up of seven specific geometric pieces that fit together to make a square. The objective of the puzzle is usually to rearrange the pieces into an image of various shapes or pictures, using all seven pieces to produce the exact shape given. For the purposes of this lesson, children discuss the different shapes shown in the tangram. However, for extension you can give children time to play with the sets of shapes.

Warm-up
- <u>Think and share</u>: Talk about the puzzle at the top of **Pupil Book 2 page 20**. Ask the children whether they have seen this game before and whether they know how it works. Some children may have their own sets or may have played with one before. If not, ask them to share their ideas and guesses about the shapes and how the puzzle works before you share the

information.
* The children may notice the seven different shapes of the puzzle: two large triangles, one medium triangle and two smaller triangles, plus a square and another 4-sided shape. They may also notice shapes formed by putting other shapes together: for example, the two large triangles together form a bigger triangle. The seven shapes together form a large square. They may also notice the irregular 4-sided shapes formed by pairs of smaller shapes.
* Discuss the names of the shapes they can see (*triangle*, *square*) and which other shapes they know that are not in the puzzle (for example, *rectangle*, *circle*) Ask whether it is possible to make, for example, a rectangle using the shapes in the tangram. Give them cut-outs or work with an online interactive tangram.

> The children are unlikely to know the name *parallelogram*, but you could ask why this 4-sided shape is not a square or a rectangle. If you wish, you can explain that we can use the word *quadrilateral* or *4-sided shape*, or simply *polygon*, to describe this shape.
>
> The children can describe shapes without using their names: for example, 'It has 3 sides and it is green.'

Focus
* For question 1, the children match the four 2D shapes to their names, in order to revise the shapes they learnt about in the previous year.
* For question 2, give each group of children a selection of 2D shapes to sort. Include rectangles, squares, triangles and circles in each set.
* For question 3, the children draw pictures made from 2D shapes. They can refer to the shapes you have given them or the shapes in the tangram.
* For question 4, the children then identify each others' shapes in their pictures from question 3. Encourage them to describe the different properties of the shapes (for example, curved outlines or 3 corners) as they explain their reasoning.

Challenge
Print several copies of tangram puzzles you find online. Cut some of the copies into the separate tangram pieces. Let the children try to construct the puzzles using the pieces. If you have online access in your classroom, children can also do this on a computer with an online interactive tangram.

Interesting mistakes
It is very likely that some children will identify the parallelogram as a rectangle, as it has two pairs of opposite sides that are equal length. This is a very useful mistake as it can help us to learn what a rectangle is. You can ask: *Why is it so easy to make this mistake?* There is something that is the same about this shape

and about a rectangle, but there is also something different. Let the children work it out.

> ## Answers for Pupil Book 2 page 20
> <u>Think and share:</u> Three different shapes.
> The children should be able to name triangle and square. (The other shape is a parallelogram.)
> Possible answers: a shape with four equal sides; a shape with three straight sides
> **1** From left to right: circle, triangle, square, rectangle
> **2** – **4** Individual's answers

Properties of shapes

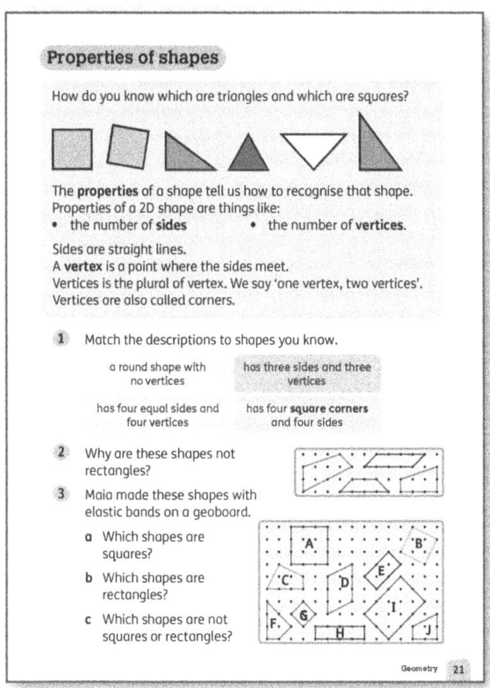

Materials
A range of objects to demonstrate each shape – aim for different dimensions of each type; flashcards with the names of the shapes; shapes (square, rectangle, triangle, circle); modelling clay; geo-strips; dotted paper; cardboard; colouring pencils; geoboard; rubber bands

Warm-up
Give the children a set of mixed 2D shapes of different sizes. Ask them to sort the shapes into groups. Give them time to verbalise the reasons for their groupings. Be aware that some children may not sort by shape at this stage and, for example, they may place all the big shapes together or all the red shapes together. Question them to lead them to other methods of sorting the shapes by characteristics related to shape, such as number of sides.

Focus
* The children could use the shapes to make patterns and designs. Encourage the children to combine

shapes to see if they can make new shapes. For example, they could put two squares together to make a rectangle.

- Ask the children to use strips, pinboards and dotted paper to construct squares, rectangles, triangles and other 2D shapes. They should count the number of sides and corners on each shape they construct. They can experiment by varying shape while keeping the number of sides the same.
- Challenge the children to examine different objects in the classroom and identify and name 2D shapes.
- Do a survey in the school environment to find, name and count shapes. Discuss which shapes are most common and let the children suggest reasons. For example, there are lots of squares and rectangles because the tiles, windows, doors, walls and other features are these shapes.
- Turn to the pictures of 2D shapes at the top of **Pupil Book 2 page 21**. Discuss how we know which shapes are squares and which are triangles. Ask:
  - *What do squares have? (4 sides and 4 corners) Are there any shapes that have 4 sides and 4 corners that are not squares? (Rectangles) What else must a square have to make it a square? (All 4 sides must be equal.)*
  - *Which triangles have a corner that looks like a square? Do all triangles have this? What do all triangles have? (3 sides and 3 vertices)*
- Discuss the terms *sides*, *vertex* and *vertices* (*corners*). Draw several shapes on the board and ask the children to identify the sides and the vertices.
- Read out the descriptions in question 1 and talk about the different shapes with the class.
- Discuss why the shapes in question 2 are not rectangles. Ask: *What do they have that rectangles also have? (4 sides) What do they have that rectangles do not have? (4 square corners)*
- Children identify which shapes in question 3 are squares, which are rectangles and are neither squares nor rectangles.

Support

Providing physical geoboards for your class and letting the children create their own squares and rectangles will help to consolidate understanding of these shapes. You can also do this with triangles of different sizes.

Interesting mistakes

At this stage, children have not learnt the definition of right angles or 90-degree angles. These angles are referred to as square corners. For this reason, parallelograms may look to them as if they should be rectangles. Similarly, they may be confused to see a square in a different orientation, for example the tipped squares in question 3. These are interesting mistakes. Encourage them to think about these questions:

- *Why is/isn't this shape a square?*
- *Why does it look like a square?*
- *Why does it not look like a square?*

Answers for Pupil Book 2 page 21

1 a round shape with no vertices: circle has three sides and three vertices: triangle has four equal sides and four vertices: square has four square corners and four sides: rectangle or square
2 They do not have four square corners.
3 a A, B and G are squares.
  b E, H and I are rectangles.
  c C, D, F and J are not squares or rectangles.

Identify 3D shapes

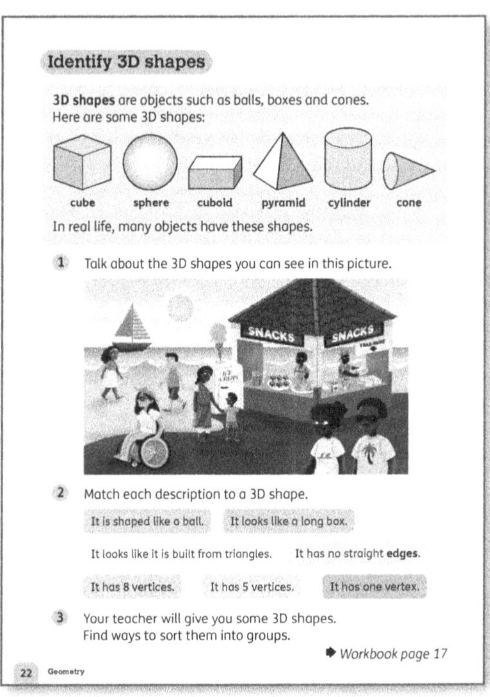

Materials

A range of objects to demonstrate each shape – aim for different dimensions of each type; flashcards with the names of the shapes; chart paper; pictures from magazines of solid objects; a feely bag with small objects representing each shape (a cotton reel, a marble, a cuboid eraser, the tip of a crayon, a cone); examples of 3D shapes (cube, cuboid, cylinder, sphere, triangular prism, sphere)

Warm-up

- Teach the names of the *3D shapes: cube, sphere, cuboid, pyramid, cylinder, cone*. Put one of each shape on display in the classroom and label it with a flashcard. Make a 3D shapes chart with names and pictures for display.
- Use the feely bag. The children take turns to put their hand into the bag and feel a shape without looking at it. They should try to guess what it is before removing it from the bag.

Focus

- Give each group a set of 3D shapes and a set of name cards. The children take turns to read a name and then find a shape to match it. The others must agree whether this is correct or incorrect. If correct, the child can use the name card to label the shape they have chosen.

- Take the children on a '3D shape walk' around the school grounds. Ask them to identify and name shapes in the environment. Use a digital camera to take photographs of the shapes they identify and then display the photographs in the classroom and check that the shapes are named correctly.
- Teach the concept of flat and curved using the 3D shapes: find out which ones roll and which do not. Spend time exploring the fact that some 3D shapes roll when placed one way, but not when placed the other way.
- The children discuss the various shapes they can identify in the picture in question 1 on **Pupil Book 2 page 22**.
- The children match each description in question 2 to a 3D shape. You can use flashcards or simply write the names of the shapes on the board.
- Put the children into groups for question 3 and give each group a selection of 3D shapes. Make sure each group has at least one of each shape to study. Ask the children to sort their shapes into groups of shapes with similar properties. Let them verbalise why they have grouped the shapes the way they have.

Follow-up

On **Workbook 2 page 17**, children draw lines to join each geometric shape to a matching real-life object.

Answers for Pupil Book 2 page 22

1 Possible answers: cuboid-shaped snack bar and ice-cream freezer; pyramid-shaped roof; ice cream cone; cylinder-shaped drink cans; pyramid-shaped snacks; cuboid-shaped boxes; sphere-shaped sun

2 It is shaped like a ball: sphere
It looks like a long box: cuboid
It looks like it is built from triangles: pyramid
It has no straight edges: sphere, cylinder or cone
It has 8 vertices: cube or cuboid
It has 5 vertices: pyramid
It has 1 vertex: cone

3 Individual's answer

Answers for Workbook 2 page 17

1 The children match A 3, B 6, C 2, D 4, E 5, F 1 and G 7.

Faces of 3D shapes

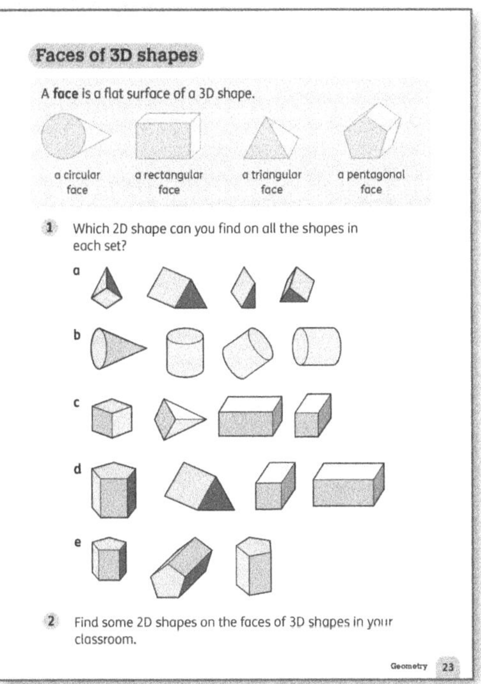

Materials

Several different boxes (for example, biscuit boxes, juice boxes); paper; scissors; sticky tack; different 3D shapes (cube, cuboid, cylinder, sphere, triangular prism, sphere); modelling clay

Warm-up

- Give each child a box. Ask them to trace one base of the box and cut it out. Tell them to draw a face on the cut-out piece. Next, get them to stick the face onto the box they were working with using sticky tack.
- Explain that the flat parts of 3D shapes are called *faces*. Discuss how many faces each box will have (six). Ask the children to trace each face, cut it out and stick it onto the box using sticky tack.
- Ask the children to remove the faces from their boxes and jumble them up. They then swap with a partner and match up the faces, re-sticking them onto the box.

Focus

- Divide the class into groups and give each group a 3D shapes. Ask the children to identify the shapes of the faces of the 3D shapes their group has. Groups should exchange shapes until they have examined them all.
- Give each child a piece of modelling clay. Ask them to make a model of a cube, a cuboid, a cylinder, a pyramid and a sphere.
- Look at the four 3D shapes at the top of **Pupil Book 2 page 23**. Discuss what each shape is and which face is shaded.
- For question 1, the children describe each set of shapes and say the 2D face that all the shapes have in common.
- For question 2, the children find everyday objects in the classroom that are 3D shapes and name the 2D shapes of the faces.

Properties of 3D shapes

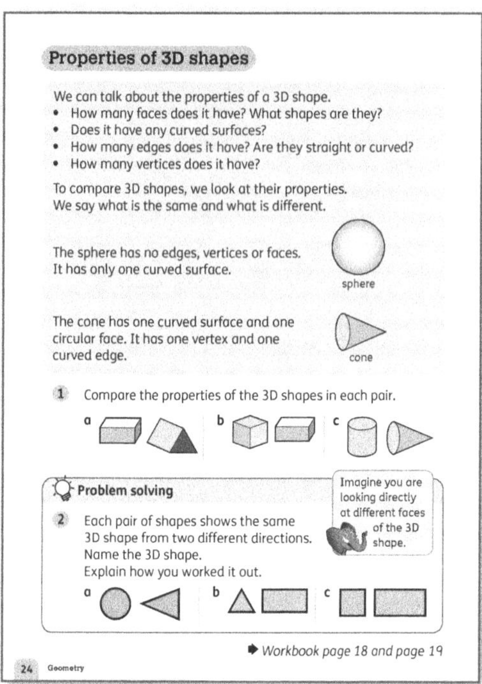

Materials
Examples of 3D shapes for the children to handle and compare

Warm-up
• Show the children two different 3D shapes and discuss what is similar and what is different about the shapes.

Focus
• Read through the information on **Pupil Book 2 page 24** to remind the children of the terms we use when we compare shapes.
• Spend some time talking about the 2D shapes we see if we hold up the shape with one face to the front. For example, hold up a cone with the circle base facing the front and ask what it looks like.
• Introduce the word *prism*. Explain that some shapes have two end faces that are the same. Show examples of a cuboid, a triangular prism, a cylinder and a pentagonal prism. Ask what the children notice about these shapes. Ask them to identify the end faces and the shapes of the end faces.
• The children compare the properties of pairs of 3D shapes in question 1.
• Problem solving: For question 2, the children need to imagine holding up a 3D shape and looking at each face directly.

Follow-up
The children complete **Workbook 2 page 18 and page 19**. Page 19 deals with the properties of 3D shapes. If you wish, you can use this as an assessment activity. Page 18 consolidates the terms *face*, *edge*, *corner*, *vertex* and *vertices*.

Challenge
Some children may like to draw a variety of prisms using different shapes for the end faces. You can introduce them to pentagons, hexagons and octagons for this. Challenge children to find everyday 2D and 3D objects and sort them into properties, for example, cans and containers of drink. Encourage them to consider the advantages and disadvantages of the shapes.

Circles

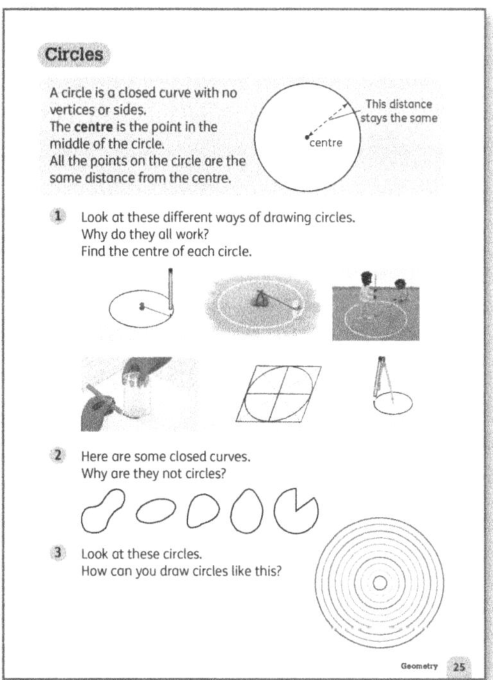

Materials

Cut-outs of circles, ovals, egg shapes and other curved shapes; pictures or examples of circular objects (sticky tape rolls, cups, wheels, plates, lids); stones; string; pieces of chalk (or other suitable materials for drawing circles outside)

Warm-up

- Say: *Imagine you meet someone who has never seen a circle. How can you explain to them what a circle is?* Let them share their ideas.
- Show the cut-outs of circles and other curved shapes. As you show each one to the class, ask: *Is this one a circle or not a circle? Why? How do you know if a shape is a circle?* Encourage them to explain their reasoning.
- Ask: *Now imagine that you need to draw a circle, but you don't have anything to draw around. How would you draw it?* Let them discuss in pairs before sharing ideas with the class.

Focus

- Take the class outside and let them work in groups with a stone, some string and a piece of chalk to draw their own circles. Take the groups around to view each other's work and talk about what worked well and what did not work.
- Turn to **Pupil Book 2 page 25**. Look at the pictures in question 1 with the class. Discuss the ways of drawing a circle that are shown in the pictures.

> One of the pictures shows a pair of compasses. Children of this age probably have not used compasses yet, but they can look at the picture and guess how they work.

- Once the children have discussed the different ways to draw a circle, revisit the questions: *What is a circle? What makes a shape a circle?* Develop the definition

that a circle traces all the points that are a fixed distance away from a *centre* point.

- The children explain why each shape in question 2 is not a circle.
- They discuss the circles in the picture in question 3 and suggest how they were drawn.

Interesting mistakes

When you deal with shapes, the circle can be confusing. It is important to remember that it is not a polygon – it is a flat shape with one curved *boundary*. Children may ask: *Can we say that a circle has one curved side?* Strictly speaking, a side is a straight line that joins two vertices, so a circle does not have sides. It has a curved boundary line or a curved outline.

> ### Answers for Pupil Book 2 page 25
> ❶ They are all made by drawing a line that stays the same distance from a fixed point. The centre is the fixed point in the middle of the circle.
> ❷ The points on each curve are not all the same distance from the centre of the shape.
> ❸ Possible answer: Use a pair of compasses. Use the same centre each time and draw the circles using different settings of the compasses.

Polygons

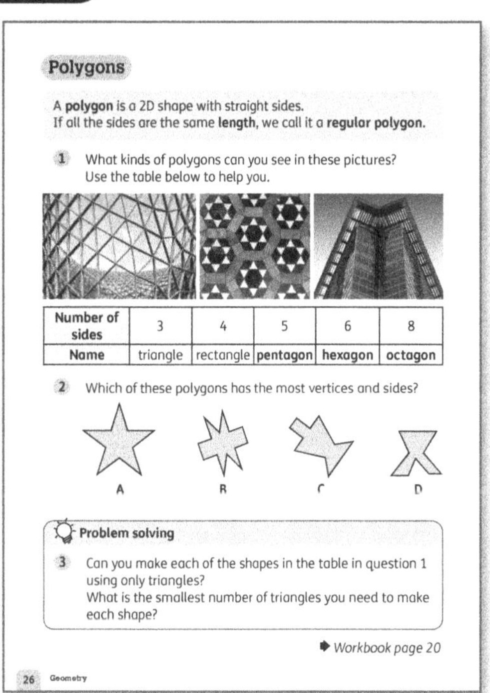

Materials

Cut-outs of some different polygons (rectangles, triangles, hexagons and pentagons – several of each); pictures of geometric patterns; a set of pattern shapes or access to an interactive set, which are readily available online

Warm-up

Show the class some examples of geometric patterns. You can use the pictures on **Pupil Book 2 page 26**, or a simple Internet search for geometric patterns will turn up thousands of examples. Many of these can be printed out as colouring sheets for additional activities if needed.

Focus

- Discuss the word *polygon* with the children. Explain that *poly* means many, and *-gon* means side. Remind them that when we talk about shapes, sides are straight lines that join up two vertices or corners. Point out that a *regular polygon* has sides the same *length*.
- Show the cut-out shapes and ask the children to help you sort them into groups with the same number of sides. As you hold up each shape, ask: *How many sides does it have?* Make four groups of shapes.
- You could use a card or sheet of paper marked with the number corresponding to the number of sides (3, 4, 5, 6).
- You could mention the names of the shapes (*triangles, rectangles, pentagons, hexagons*) and relate these to the number of sides. However, at this stage, focus on sorting and comparing the polygons, rather than the shape names.
- Discuss the polygons the children can see in the pictures in question 1.
- For question 2, children count the vertices and sides of the polygons to work out which has the most.
- Problem solving: For question 3, the children draw triangles to make up rectangles, pentagons, hexagons and octagons. If required, provide cut-outs of these shapes and let the children draw diagonals to divide the shapes into triangles.

Follow-up

The children complete **Workbook 2 page 20** about the sides and corners of polygons.

Challenge

If you have a set of pattern blocks, let the children use these to make up different polygons. You can also find interactive versions online.

Support

The Challenge suggestion above using pattern blocks is equally suitable for children who need support as it is for those looking for a challenge. You can also give children geoboards and elastics, or dotted paper, and let them practise drawing polygons with given numbers of sides.

Answers for Pupil Book 2 page 26

1 1st picture: triangles, hexagons (made from 6 triangles)

2nd picture: triangles, squares, rectangles, hexagons

3rd picture: rectangles, triangles

2 B

3 Problem solving: triangle:

1 triangle    rectangle: 2 triangles

pentagon: 3 triangles

hexagon: 4 triangles

octagon: 6 triangles 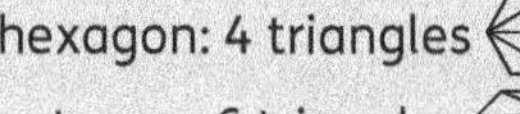

Answers for Workbook 2 page 20

1 **a** Individual's own drawings of two different squares; 4 equal sides, 4 vertices

b Individual's own drawings of two different triangles; 3 sides, 3 vertices

c Individual's own drawings of two different rectangles; 4 sides, 4 vertices

End-of-unit check

Ask questions and carry out practical activities to assess the children's understanding of 2D shapes, for example:

- *What shape is this* (show a 2D shape)*? How do you know?* (For example, the child counts the number of sides or vertices and uses this to identify the shape.)
- *What shapes have 4 sides?* (rectangles and squares)
- *What shapes are round?* (circles)
- *What shape has 3 sides?* (triangle)
- *What is the same about a rectangle and a square?* (Both have 4 sides and square corners.)
- *What is different about a rectangle and a square?* (A square has 4 equal sides and a rectangle has 2 pairs of sides that are equal.)
- Show a variety of curved shapes and circles and ask the children to explain which are circles and which are not. Ask: *How do you know which ones are circles?* (All points on a circle are the same distance from the centre point.)
- Show a variety of polygons and ask the children to say how many vertices and sides they have. Alternatively, provide cut-outs of polygons and ask the children to sort them from least to most vertices/sides.

Assess the children's understanding of 3D shapes by repeating the following activities from class:

- Use a feely bag (page 23). Let the children take turns to put their hand into the bag and feel a shape without looking at it. They should try to guess what it is before removing it from the bag. (Encourage the children to identify the if the 3D shapes have curved surfaces or flat faces and the number of faces, edges and vertices and to use this information to name them.)
- Give each group a set of 3D shapes and a set of name cards. The children take turns to read a name and then find a 3D shapes to match it. The others must agree whether this is correct or incorrect. If correct, the child can use the name card to label the shape they have chosen. (For example, a child has 'cube' on a card and finds a cuboid; the other children indicate that this is incorrect.)
- Provide some different 3D shapes and ask questions about the faces, vertices and edges.
- Give the children a selection of 3D shapes and ask them to identify one that matches a given description. For example, say:
  - *Find me a 3D shape that has 1 circular face and a curved surface.* (cone)
  - *Find me a 3D shape that has 6 square faces.* (cube)
  - *Find me two different 3D shapes that have triangular faces.* (triangular prism and pyramid)

UNIT 5

Patterns and sequences

Learning objectives

- Recognise, describe and extend numerical sequences (0 to 100).
- Order and arrange combinations of mathematical objects in patterns and sequences.
- Count in steps of 2, 3 and 5 from 0, and in tens from any number forward and backward.

Key words

pattern shape repeat repeating shapes growing patterns core terms rule

Unit introduction

Materials

Number cards (optional)

Teaching guidance

Use a number talk to open this unit (see 'Number talks', pages 17–18). Here is a problem you can start with:

- Use number cards or draw on the board to set up the following number puzzle:

6	7	8	
		18	
	27	28	29

- Ask the children to look at the number puzzle and try to work out what the *pattern* is. Let them think about it individually for a while. After everyone has indicated that they have some ideas, allow the children time to share ideas quietly in pairs.
- Then say that you are going to show a number card and they can put their hands up if they know where it goes. Hold up the number 39. Invite a child to come up to place it. Repeat this with the numbers 40, 50, and so on.
- Invite children to come up and place numbers in the correct place. Gradually they should establish that the numbers form part of the 100 chart. However, they may explain this in different ways. For example: from left to right, the numbers go up by 1 and as they go down, they go up by 10. Let the children gradually extend the puzzle in various directions, for example:

6	7	8		
		18		
	27	28	29	
		38	39	40
				50
				60
				70

For more problems that you can use in number talks on this topic, use 'Number sequences' and 'Function machines' (page 30).

Look at patterns

UNIT 5 Patterns and sequences

Look at patterns

Think and share

What do you notice about all these pictures?

Which real-life objects have **patterns** that people have made?
Which objects have natural patterns?
Which shapes do you notice?
Where do you see patterns in real life?

We can make patterns out of shapes, objects and numbers.

Workbook page 21

Number 27

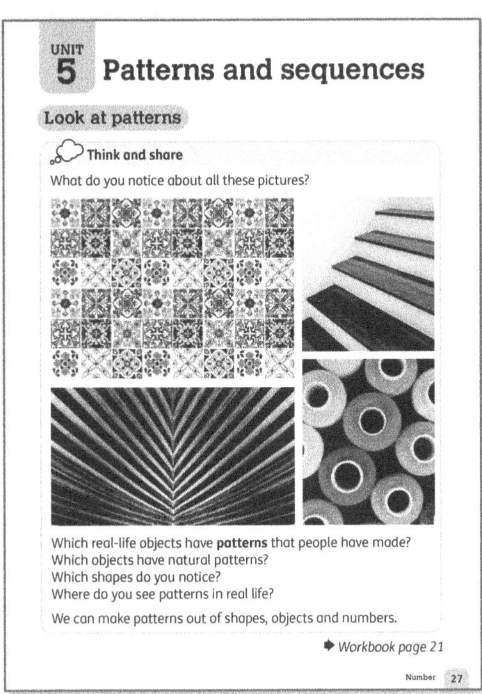

Materials

Pictures of a variety of patterns, such as patterns in nature (honeycomb, shells, wave and sand patterns, flowers, leaves, patterns created by crop rows); fabric patterns; tiling patterns; architectural patterns

A unit on patterning can be a huge amount of fun for children if you ask them to pay attention to the wide range of patterns that are all around us in the natural and built world. Building up a collection of pictures with a variety of patterns is strongly recommended. You can find a multitude of resources online. Try searching for 'natural patterns', 'aesthetic patterns', 'art patterns', 'drawing patterns' or 'geometric patterns'.

Teaching guidance

Focus

<u>Think and share:</u> Discuss the pictures on **Pupil Book 2 page 27** with the class. Ask questions to guide their understanding, for example:

- *What can you see?*
- *What makes the pattern?*
- *Is the pattern mainly made by shapes, or colours or both?*
- *Where would you see patterns like this in real life?*
- *What other patterns do you see in nature?*

If you have more pictures to show, extend the discussion to talk about other patterns.

Discuss with the class what we mean by pattern and *repeat*, including *repeating shapes*. Often, patterns are created by repeating something over and over in a regular way.

Follow-up

The children can work through **Workbook 2 page 21**. They complete each pattern by repeating the same shapes or types of line. They can also colour their designs.

Answers for Pupil Book 2 page 27

<u>Think and share:</u>
Possible answer: The pictures here are all of patterns. They show: decorative tiles, steps on a staircase, a palm leaf, spools of cloth or yarn/thread. We can see shapes such as circles, lines, rectangles and squares.

Answers for Workbook 2 page 21

1 **a–d** Individual's completed patterns using similar shapes

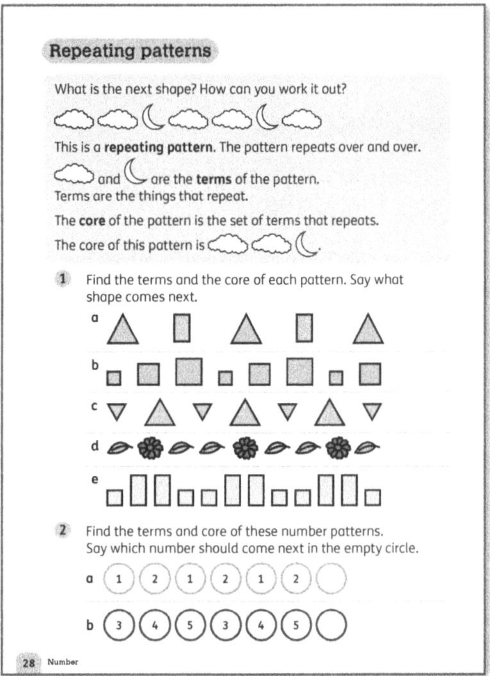

Materials

Cut-outs of various shapes or pictures that can be used to make repeating patterns; sticky tack or some other way of pinning up shapes (such as a washing line and pegs); pattern shapes (optional)

Warm-up

Start with a number talk (see 'Number talks', pages 17–18). Explain to the children that you are going to put up some shapes, and they need to guess which one comes next. One at a time, put up shapes to create a simple repeating pattern.

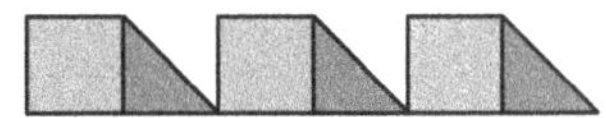

Ask the children to predict what will come next. Let them explain how they worked it out. Repeat with this pattern:

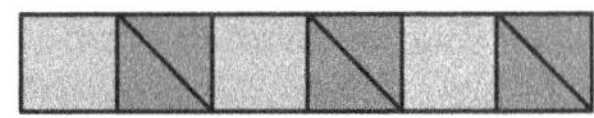

Here is another pattern, with less familiar shapes:

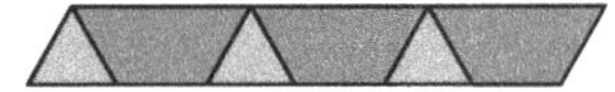

Let the children suggest what to add each time. You can introduce the term 'trapezium'.

Focus

Talk about the pattern at the top of **Pupil Book 2 page 28**. Use this to introduce the words *terms* and *core*. Ask the children what the shapes are that make the pattern. Ask: *How many shapes are there? How many times do they repeat? What would the next shape in the pattern be?* (The next shape would be a cloud, because the pattern goes cloud–cloud–moon.)

- Let the children work in pairs to complete question 1, encouraging them to discuss their ideas and reasoning with their partner as they complete the patterns.
- Let the children complete question 2 individually, assisting as needed.

Challenge

Challenge the children with patterns of unfamiliar shapes or patterns made by turning the shapes in different directions:

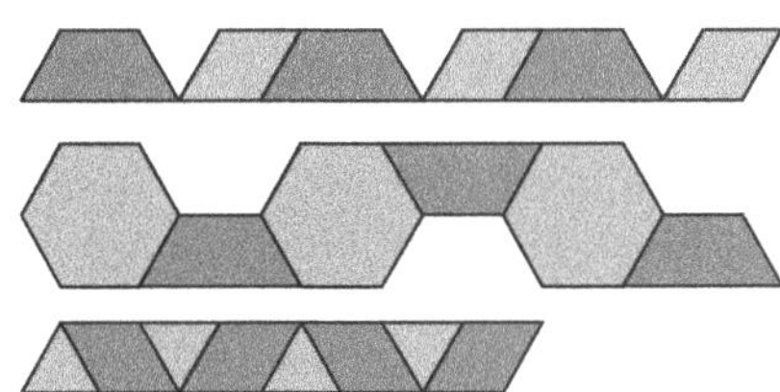

For extra challenge, use a physical set of pattern blocks or virtual, online versions for children to create their own more complex patterns.

Support

Pattern shapes are a wonderful resource for creating simple repeating patterns using geometric shapes. Use a set of wooden or plastic pattern shapes or use an interactive online set.

Answers for Pupil Book 2 page 28

1 **a** terms: triangle and rectangle
core: triangle rectangle
next shape: rectangle
b terms: small square, medium square and large square
core: small square medium square large square
next shape: large square
c terms: small triangle and large triangle
core: small triangle large triangle
next shape: large triangle
d terms: leaf and flower
core: leaf flower leaf
next shape: leaf
e terms: square and rectangle
core: square rectangle rectangle square
next shape: square
2 **a** terms: 1 and 2   core: 1 2   next number: 1
b terms: 3, 4 and 5   core: 3 4 5   next number: 3

Find the rule

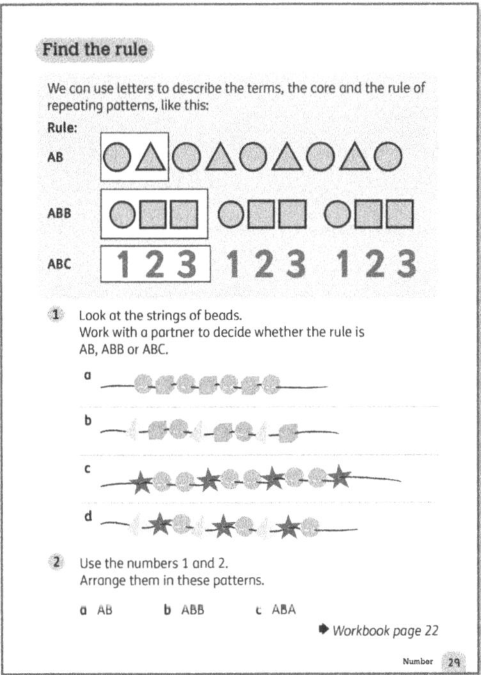

Warm-up

Return to the patterns you made in the previous lesson. Explain to the children that the shapes we use to make a pattern are called the *terms* of the pattern. Ask them to identify the terms in some of the patterns they worked with before.

Focus

- Demonstrate how we can give the terms letters to describe the way the pattern repeats. You can do this using the patterns from the previous lesson or using the material on **Pupil Book 2 page 29**.
- Look at each string of beads in question 1 with the class. Discuss the terms of each pattern (the shapes that make up the pattern), as well as the *rule* (how the terms repeat). In many cases the core of the pattern is simply made up of the terms arranged in a repeating sequence. However, in some cases one term is repeated more than once within the core.
- Discuss with the children how to create the patterns in question 2 using numbers.

Follow-up

On **Workbook 2 page 22**, the children label the terms of the pattern A and B, and then identify the core of each pattern. Then they make their own patterns.

Answers for Pupil Book 2 page 29

1 **a** AB   **b** ABC   **c** ABB   **d** ABC

2 **a** 1 2 1 2 1 2 …   **b** 1 2 2 1 2 2 …

    c 1 2 1 1 2 1 …

Answers for Workbook 2 page 22

1 **a** AB   **b** ABA   **c** ABB

2 Individual's answers. The children's patterns should be made up of repeats of their core shapes, with no extra shapes and no shapes missing.

Repeating shapes and numbers

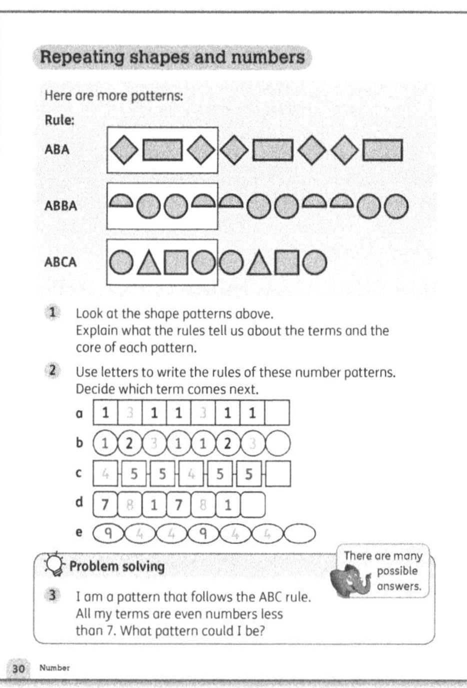

This lesson continues from the previous work identifying the rule of repeating patterns. The children should now begin to notice more complex core units.

Materials
Wooden/plastic shapes or shape cards

Warm-up
Start the lesson by reviewing the patterns from the previous lesson.

Focus
For question 1, discuss the patterns at the top of **Pupil Book 2 page 30** with the class. Ask:

- *What shapes make this pattern?* (Rhombus or tilted squares and rectangles/ semicircles and circles/ circles, triangles and squares)
- *What are the terms?*

- *How do they repeat?*
- *Who can see the core of this pattern?*
- *What does ABA tell us? Which shape is A? Which shape is B?*
- *What does ABBA tell us? How many shapes are in the core of this pattern?*
- *What does ABCA tell us? Why is there a letter C in this pattern?*

The children inspect each pattern in question 2 and identify the pattern rule using the letters.

Problem solving: The children use the information in question 3 to work out possible patterns. Remind them that the pattern does not tell us anything about the order, so it is possible to arrange the terms in any order as long as they follow the ABC rule.

Support
Some children may confuse the letters of the rules with the terms themselves. It is important to give them plenty of practice using either concrete shapes or shape cards and making simple ABAB and ABB patterns so that they can learn how to identify the terms and the rule. Do this before moving on to more complex repeating patterns.

Answers for Pupil Book 2 page 30

1 The letters tell us how the terms repeat.

ABA tells us there are two terms: A and B. They repeat in the core pattern ABA (square rectangle square).

ABBA tells us there are two terms: A and B. They repeat in the core pattern ABBA (semicircle circle circle semicircle).

ABCA tells us there are three terms: A, B and C. They repeat in the core pattern ABCA (circle triangle square circle).

2 **a** rule: ABA   next term: 3

    b rule: ABCA   next term: 1

    c rule: ABB   next term: 4

    d rule: ABC   next term: 7

    e rule: ABB   next term: 9

Problem solving:

3 The even numbers less than 7 are: 2, 4 and 6. All the possible answers are: 2 4 6 2 4 6 …; 2 6 4 2 6 4 …; 4 2 6 4 2 6 …; 4 6 2 4 6 2 …; 6 2 4 6 2 4 …; 6 4 2 6 4 2 …

[Note that patterns that include repeated numbers are not correct because they don't follow the ABC rule.]

Growing patterns

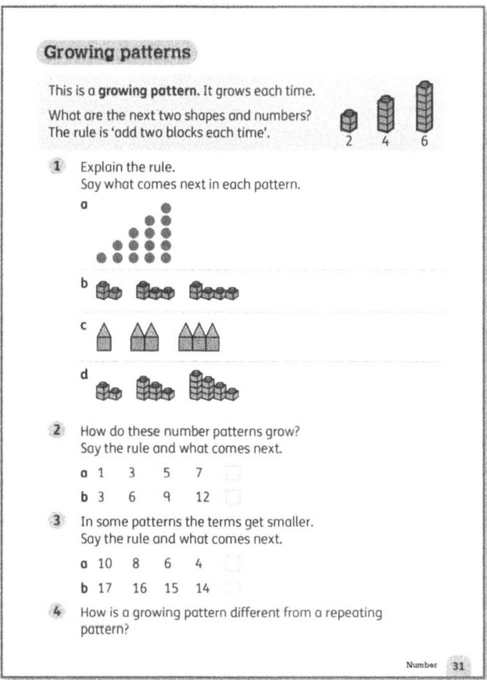

Materials
Interlocking cubes (page 21), counters and other materials for building growing patterns

Warm-up
- Use concrete materials to set out a pattern like the one shown at the top of **Pupil Book 2 page 31**. Set out the terms one by one. Ask:
  - *What pattern am I making?*
  - *What do you think comes next? How did you work it out?*
  - *How can you describe this pattern?*
  - *Can you think of another pattern we can make like this?*
- Let the children share their ideas. They may notice that the shape increases by 2 cubes each time and that the numbers grow in the same way – we add 2 each time. Some may have different ways of expressing this. They may notice that on a number line this pattern is the same as skip counting in twos. Or they may notice that it is the same as counting every second number on the 100 chart. Ask: *What are the next two towers and numbers?* (The next towers have 8 and 10 cubes respectively, and the numbers are 8 and 10.)
- You may need to follow this strategy (using concrete materials and conducting a small number talk) with each pattern in questions 2 and 3.

Focus
For question 1, ask questions to guide the children to express what they see.

The pattern in part a is a simple *growing pattern*. Ask questions such as:

- *This pattern starts with 1 counter. What happens in the next step?* (It goes up by 1.)

- *How many more counters do you add each time?* (1)
- *What shape does it make?* (a triangle)
- *What do you think will come next?* (a column of 6 counters)

The pattern in part b is also a growing pattern, but it grows in a different shape. Ask:

- *How many cubes make the first shape?*
- *What is different between the first and the second shape?*
- *What is different between the second and the third shape?*
- *What gets added on each time?*
- *How would you build the next shape?*

Continue in the same way with the patterns in part c and part d.

- By now, the children should be used to comparing each term in a sequence to work out the rule. Number patterns work in the same way.

For each of the number patterns in question 2, the children need to work out how much to add to get the next number.

- If you have time, you can give them a variety of materials (such as counters, cubes or sticks) and let them build their own shape patterns using the relevant number of objects for each shape term.
- They can explore different ways of showing the same sequence using objects (for example, 1 stick, 3 sticks, 5 sticks, and so on).
- They also need to be able to examine a number sequence and work out how it changes from one term to the next.

For question 3 and question 4, the children may want to discuss the concept of growing patterns that grow smaller.

- The key here is that while repeating patterns repeat the same unit of pattern over and over, growing patterns change using a regular addition or subtraction.
- At higher levels, children may come to understand this as addition of a positive or negative number. At this stage, you can explain that in the same way that we can add on a number again and again, we can also take a number away to make the next term.

Challenge
Invite the children to make their own growing patterns using cubes or counters. A partner must explain the rule and decide whether the pattern makes sense.

Support
For children who need additional support to express the 'rule' of a pattern, let them focus on making visual sense from a pattern with concrete materials. Ask them to explain in their own words why they think their pattern 'works'.

Answers for Pupil Book 2 page 31

1 a rule: add one circle each time; next: a column of 6 circles

b rule: add one cube each time; next: 6 cubes (arranged as a 2 and 4 ones)

c rule: add one 'house' (or one square and triangle) each time; next: a row of 4 'houses' (or 4 squares and 4 triangles)

d rule: add a tower with one more cube each time; next: 5 towers of 5, 4, 3, 2 and 1

or rule: each time, add one cube to the top of each tower and then one more cube; next: 5 towers of 5, 4, 3, 2 and 1

2 a add 2 each time **or** skip count odd numbers; next: 9

b add 3 each time **or** skip count in threes; next: 15

3 a take away 2 each time **or** skip count back in twos; next: 2

b take away 1 each time **or** count back in ones; next: 13

4 Possible answer: In repeating patterns the same terms are repeated in the same arrangement. In growing patterns, the terms are all different/keep changing. The terms get greater or smaller.

Skip-counting patterns

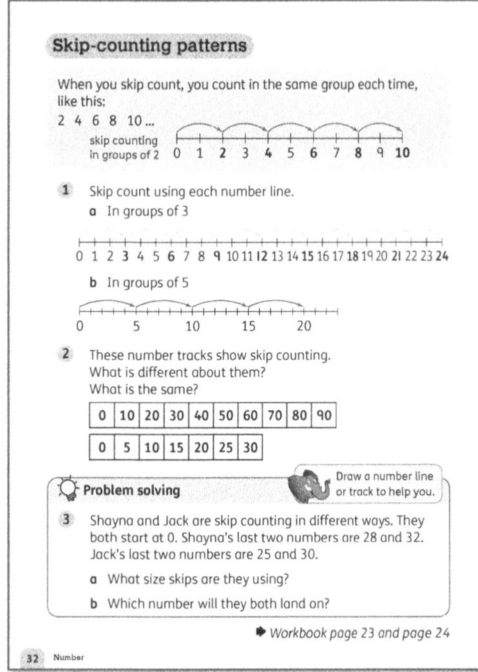

Materials

100 chart; number lines; number tracks; number cards (0–20)

A number line:

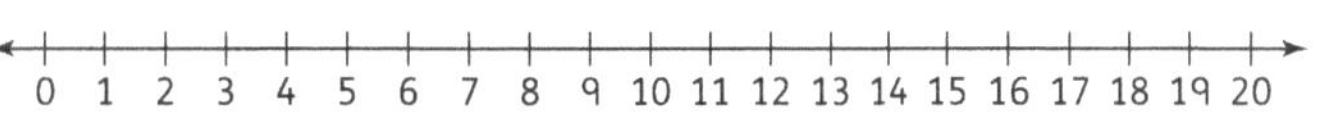

A number track:

1	2	3	4	5	6	7	8	9	10	11	12	13	14	15	16	17	18	19	20

Warm-up

- Display a 0–20 number line on the board. Count with the class from 0 to 20. Ask a child to come up and point to the numbers you counted in order.
- Then ask the class to listen to a different set of numbers. Count in mixed-up order, such as: 7, 10, 9, 4, 2, 3, 8, 10, 1. Ask: *Does anyone know which way I was counting? Why was it hard to follow?*
- Say: *Now listen carefully. I want you to explain how I've counted.* Count 0, 2, 4, 6 … up to 20. Ask whether someone can show on the number line which numbers you counted. Ask: *Are you sure those were the numbers I said? Why is it easy to remember this time? What was the pattern?*

Focus

- Turn to **Pupil Book 2 page 32**. Skip count in twos pointing to the correct numbers on the number line.
- For question 1, skip count with the children: in threes for part a and in fives for part b.
- Ask the children to look at the number tracks in question 2 and say what is different and what is the same.
- <u>Problem solving:</u> For question 3, the children have to work out the size of the skips each child used in their count, and then find the numbers the two sequences have in common.
- They need to start by counting on from 28 to 32 to find that there are 4 jumps to the next number. Then they can work backward to make the rest of Shayna's sequence. If you draw a number track with 10 spaces, they can write the last two numbers in, and work backward:

								28	32

- It should be easier for the children to work out Jack's sequence as they have just been looking at skips of 5.

Follow-up

The children complete **Workbook 2 page 23 and page 24** for extra practice with skip counting and number patterns.

Support

For children who need extra practice with skip counting, set out sequences of number cards from 0 to 20, and turn over the numbers you are skipping so that only the multiples remain face up.

Challenge

Invite the children to make skip-counting patterns by starting with any number between 0 and 50 and adding on 10 each time. Ask: *What is the same with each term in the number pattern?* (The ones digit stays the same.) *What is different for each term in the number pattern?*

(The tens digit changes by 1 each time). The children could also create skip-counting patterns by starting with any number between 50 and 99 and counting back by 10 each time.

Answers for Pupil Book 2 page 32

1 a 3, 6, 9, 12, 15, 18, 21, 24
   b 5, 10, 15, 20

2 Possible answer: Differences: The top track shows skip counting in groups of 10 and the bottom track shows skip counting in groups of 5. Similarities: They both go up by the same number each time. They both start from 0. They both include the numbers 0, 10, 20 and 30.

<u>Problem solving:</u>

3 a Shayna: 4; Jack: 5   b 20

Answers for Workbook 2 page 23

1 a 25, 30, 35, **40, 45, 50**   Skip count on in 5s
   b 40, 50, 60, **70, 80, 90**   Skip count on in 10s
   c 40, 38, 36, **34, 32, 30**   Skip count back in 2s
   d 43, 46, 49, **52, 55, 58**   Skip count on in 3s
   e 98, 87, 76, **65, 54, 43**   Skip count back in 11s
   f 11, 22, 33, **44, 55, 66**   Skip count on in 11s
   g 12, 23, 34, **45, 56, 67**   Skip count on in 11s
   h 56, 52, 48, **44, 40, 36**   Skip count back in 4s

Answers for Workbook 2 page 24

1 a 30, 35, **40**, 45, **50, 55**, 60, **65, 70**, 76
   Skip count on in 5s
   b 36, 46, **56, 66, 76, 86**, 96
   Skip count on in 10s
   c 23, 33, 43, **53, 63, 73, 83**
   Skip count on in 10s
   d 47, 50, 53, **56, 59, 62, 65, 68, 71**
   Skip count on in 3s
   e **64, 67, 70, 73, 76**, 79, 82, 85
   Skip count on in 3s

2 a 46, 48, 50, **52, 54, 56**
   b **85, 80, 75**, 70, 65, 60
   c **44, 49, 54**, 59, 64, 69
   d 43, 47, 51, **55, 59, 63**

End-of-unit check

Carry out activities to assess the children's understanding of patterns, for example:

- Show pictures of patterns (such as leaf patterns, flower patterns, wave patterns, honeycomb) and ask them to describe what different shapes and patterns they can see.
- Create simple repeating patterns using shapes or objects and ask the children to continue by placing the next shape or object.
- Set out a repeating pattern and ask the children to identify the core (the part that repeats).
- Follow a similar procedure with numeric patterns, for example:

  5 4 5 4 5 4 …

  3 2 2 3 2 2 3 2 2 …

  6 7 6 6 7 6 6 7 6 …

  1 4 4 1 1 4 4 1 …

- Set out simple growing patterns using counters or cubes. Ask the children to say whether each one is a growing or repeating pattern, and how they know. (For example, the child indicates that for a growing pattern the numbers are increasing and the child explains that for a repeating pattern the same numbers repeat in a sequence). Ask them to set out each next term in each pattern.
- Ask the children to skip count in a given number (such as 2, 3, 5 or 10). They can use a number line, number track or 100 chart to help them.

Add and subtract

Learning objectives

- Recall and use addition and subtraction facts to 20 fluently and derive and use related facts up to 100.

- Solve problems with addition and subtraction: using concrete objects and pictorial representations, including those involving numbers, quantities and measures; applying their increasing knowledge of mental and written methods.

- Add and subtract using concrete objects, pictorial representations and mentally: 2-digit numbers and ones; 2-digit numbers and tens; two 2-digit numbers; adding three 1-digit numbers.

- Show that addition of numbers can be done in any order (commutative), but that subtraction cannot.

- Recognise and use the inverse relationship between addition and subtraction and use this to check calculations and solve missing number problems.

- Count in steps of 2, 3, and 5 from 0, and in tens from any number, forward and backward.

- Recognise the place value of each digit in a 2-digit number (tens, ones).

- Identify, represent and estimate numbers using different representations, including the number line.

Key words

add   subtract   total   sum   order   count on   pair   round

Unit introduction

Materials

Number cards (0–5); small containers, such as bowls, bottle lids; small stones or counters; number lines

Teaching guidance

This unit introduction starts with a number talk (see 'Number talks', pages 17–18). It then suggests a range of practical activities you can use for addition and subtraction work. You may choose to use some, all or none of these activities. However, it is strongly suggested that you use the number talk as a way of keeping children discussing their reasoning and practising their problem-solving skills.

- Start with a talk about a numberless word problem (see 'Number talks' and 'Numberless word problems', pages 17–18). Remember, the aim of a number talk is to get children to engage in conversations about maths, not to get the right answer as fast as possible. Suggestions and questions are encouraged, as are mistakes, which offer opportunities for discussion.

- Write on the board and say, for example: *I have some pens. I get some more pens. Now I have lots of pens.* Then ask:
  - *What questions can I ask about this story?*
  - *What would you like to find out?*

- During discussions, check the children's understanding of the terms *total* and *sum*.

- Cross out 'some more' and write 10 in its place. Ask: *Now what do we know about the number of pens that I start with?* (We don't know how many.) *Do we know anything about the number of pens I will end with?* (It's 10 more than you started with.) Say: *OK, I'm going to give you a question: How many pens did I start with?*

- *What information do I need in order to work it out?*

- Cross out 'lots of' in the third sentence and replace it with the number 23. Say: *Now I have more information. Who has an idea how we can work this out?*

- Continue in the same fashion as you do with other number talks.

- Give each group two small containers and 10 stones or counters. Working in pairs, the children must take turns to take a few stones and put them into one of the containers, so there are stones in each container. Explain that they are going to *add* them to find out how many there are. The children must discuss how to do this and find ways of working out what the *total* (*sum*) is. Get them to check by counting the stones one by one.

- Give each child a 0–10 number line. Play games where you tell them the starting point and how many places to move. They should shout out where they will land and then make the moves to check that they are right. Repeat this several times with many different starting points, including 0. You can also say *Move 0 places* to reinforce the concept of adding 0.

- Give each group a set of 0 to 5 number cards, a pile of stones and two containers. They must choose two number cards without looking. They should then make the numbers with two groups of stones and say how many they have in total.

- For more opener activities and extra practice activities that consolidate understanding use: 'Building numbers in place-value tables' (page 27); 'Target diagrams' (page 29); 'Target number calculations' (page 29) and 'Target number instructions' (page 30).

Bar models

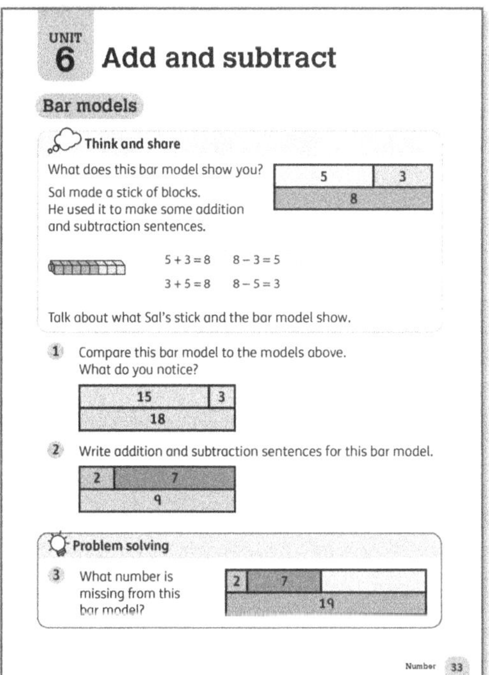

Materials

Interlocking cubes (page 21)

Warm-up

Bar models (see page 22) are a powerful way of
representing the relationship between addition and
subtraction. Begin a number talk with a discussion about
bar models (see 'Number talks', pages 17–18). Show a
few examples without any explanation, for example:

10	
7	3

10	
4	6

When you draw the bar models, your proportions
do not have to be exact, but they should bear
some logical resemblance to the numbers you
are representing. So, if you are showing 2 fives are
equivalent to 10, the second bar should be split
around the midpoint.

Ask:

* *What can you see?*
* *What do you think this picture is showing us?*
* *What could it help us to work out?*

Let the children suggest ideas. They may make suggestions
such as: 'It shows that seven add three equals ten.'

Ask:

* *Is there any other way we could say this?*
* *What about taking away? What does it tell us we can
take away from 10?*

* *What would be left?*
* *How can I fill in the empty bars in this bar model?*

10	

Again, invite the children to share their ideas. Some may
suggest other number bonds to 10, such as 1 and 9 or 8
and 2; others may notice the equivalent size of the bars
and suggest 5 + 5.

Focus

* Explain that we call these diagrams *bar models* and
that we can use them to help us understand adding
and taking away. Explain that adding is also called
addition. It means putting parts together to make a
larger group or number.
* As you work through example bar models, emphasise
that the models show us that we can break a number
into parts. Explain that taking away a part of a
number is called subtracting. Up until now we have
called it taking away, but children will start to see
the words *subtract* and *subtraction*. Relate this to the
minus sign. Show how we can represent the related
number facts for a bar model:

$$10 = 8 + 2 \quad 10 = 2 + 8$$
$$10 - 8 = 2 \quad 10 - 2 = 8$$

* It is imperative that children begin to model the
idea of part + part = whole. Bar models help them
to think of numbers in this way, and to recognise
the commutative property (i.e., whether the *order*
matters) of addition and subtraction:
whole − part = remaining part.
* <u>Think and share:</u> Turn to **Pupil Book 2 page 33**.
Discuss Sal's bar model with the class. Ask what he
did with the cubes. It is important to give children
opportunity to build their own bar models with
interlocking cubes in this way so that they can see
the actual relationship between the quantities, not
just the numbers, which could be incorrectly filled in.
* For question 1, let the children consider the similarity
between the bar models shown. They may notice
that, in both cases, we add the ones (5 and 3) to get
a total with 8 in the ones place, although the second
model has an extra ten.
* Talk about the fact that the whole bar of the model in
question 2 is the same size as the one in question 1,
but represents a different value.
* <u>Problem solving:</u> In question 3, the top bar has been
partitioned into three parts. Ask the children whether
they can notice any similarities with the model in
question 2. Again, they may notice that the ones add
up in the same way.

Challenge

Invite the children to use number bonds to 10 to derive
related facts to 100 and represent these number
sentences using bar models, for example, if 4 + 6 = 10
they can derive that 40 + 60 = 100.

Think and share: It shows 5 + 3 = 8.

The stick and bar model show that 5 cubes and 3 cubes are the same length as 8 cubes.

1 Possible answer: The ones digits are the same. There is another ten added to the 5, and another ten added to the 8.

2 2 + 7 = 9   7 + 2 = 9   9 – 2 = 7   9 – 7 = 2

Problem solving:

3 10

Add and subtract

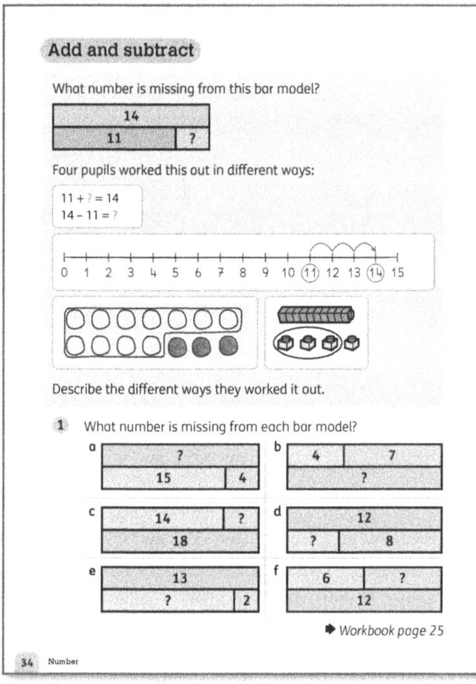

Materials

Interlocking cubes (page 21) or counters; rekenreks (page 21)

Warm-up

Discuss the example at the top of **Pupil Book 2 page 34** and ask the children to describe the different ways the pupils worked out the answer to the question:

- The first child expressed the question as addition and subtraction sentences.
- Remind the children of the term *count on* when you explain that the second used a number line to count on from 11.
- The third drew a 7 × 7 array of circles to show 14 and then circled 11 circles to see how many were left.
- The fourth used interlocking cubes to make a tower of 10 cubes and then 4 single cubes to show 14.

Focus

Complete question 1 with the class. Follow the number talk structure ('Number talks', pages 17–18). Ask the children to work out the missing number from each bar model.

Follow-up

Workbook 2 page 25 gives children practice in using bar models to calculate missing numbers. They use reasoning to find the missing number, deciding whether to use addition or subtraction. They can use any strategy: counting on, counting back, using number lines or counters and interlocking cubes. Encourage the use of concrete manipulatives (counters and cubes) as needed.

Question 2 draws children's attention to the relationship between the parts of the bars. Work through the completed example with them. Ask guiding questions, such as:

- *Which part of the bar model has this pattern?*
- *What patterns do the other parts have?*
- *Which ones together make the one with this pattern?*
- *So which bit do we take away/add to get this bit?*

Challenge

You can provide some bar models with a deliberate mistake in one of the bars and ask the children to show a range of different ways to correct it. For example:

12	
4	7

Possible corrections:

12	
4	8

11	
4	7

12	
5	7

Support

Some children may need to continue working with counters and cubes as they work with bar models. Encourage this use of concrete manipulatives to keep developing their number sense. You may also want to provide children with rekenreks to model bars.

Interesting mistakes

Children may assume that the length of the bar indicates the size of the number. However, in each question it is only the relationship between the bars that line up with each other that indicates the size of each total.

They also may make the mistake of assuming that all bars should add up to a given amount (such as 10 or 20) if they have just worked with that number. Invite them to question:

- *What made this an easy mistake to make?*
- *What shows you that this is a mistake? How can we correct it?*

Make 20

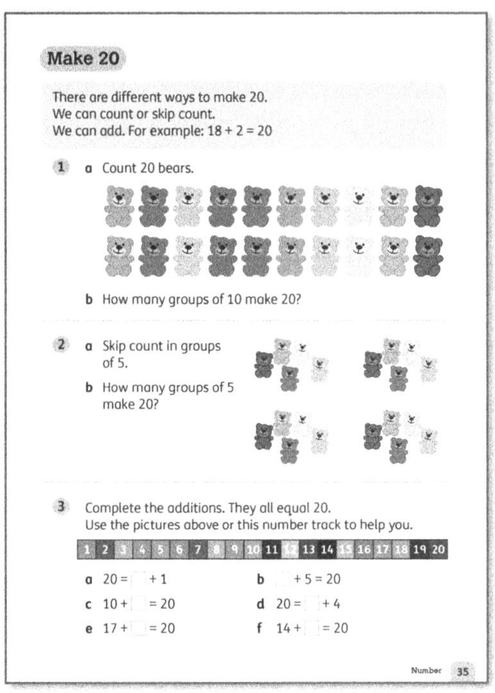

Materials
Small container such as bottle lids; small stones or counters; interlocking cubes (page 21); counters

Warm-up
Give each group two lids and 20 stones. Ask them to count the stones and then to check their counting. Ask:

- *What is an easier way to count out the 20 stones?*
- *Did anyone use groups?*

Invite the children to demonstrate some ways. For example, they might count in twos, or use 4 groups of 5 or 2 groups of 10. Ask: *So are we are all sure that we have a total of 20 now?*

Focus
- Ask one child to choose any number between 5 and 20 and to count out that number of stones into one lid. For example, a child may choose 11. Then they put the remaining stones into the other lid without counting. Ask: *Who can guess how many are in the other lid? How can we work it out?* Give time for them to think about it, work it out and then check by counting out the stones.
- Repeat this activity with a total less than 20.
- If you have coloured interlocking cubes, you can get another child to model the number bond (for example, 11 + 9) using a different colour to show each part. You can also then split the cubes into 2 tens. The colours will show clearly the partitioning as 10 + 9 + 1. Ask questions to guide children to see how we can add up 2 tens to make 20.
- You can work through questions 1–3 on **Pupil Book 2 page 35** with the children or they can work independently.

Interesting mistakes
- Some children may be confused by missing number addition sentences with the equals sign in different places. Deliberately vary the placement of the equals sign (sometimes with the total at the front of the sentence, sometimes at the end), in order to cultivate the reasoning around the significance of this sign.
- The equals sign is not an operator instructing us to find the answer – this is a mistake that children may make if we always present missing number additions with the answer after the equals sign.
- Instead, it is important to emphasise that the equals sign asks us to think about what would be equal to the expressions on each side of that sign. In Unit 8, they will revisit balance scales in the work on mass, and this is another good opportunity to reinforce the idea of equals.

Add or subtract ones

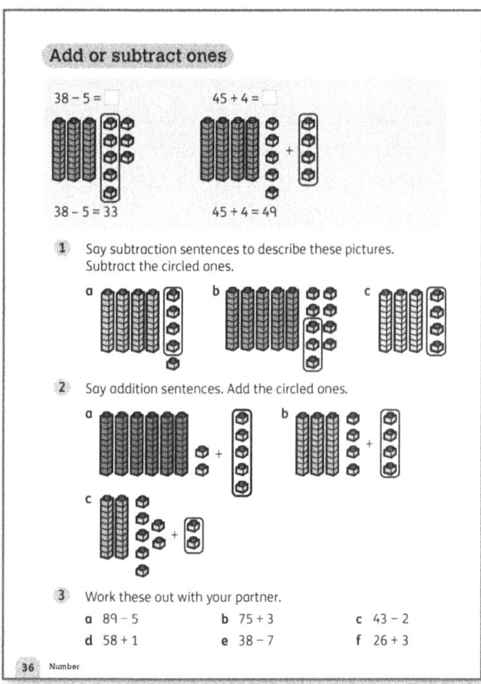

Materials
Interlocking cubes (page 21); large 100 chart, plus laminated versions for the children

> Concrete materials are invaluable for addition work with 2-digit numbers as they support children to visualise adding tens and ones. Start slowly, with additions of single-digit numbers to 2-digit numbers. At this level, children do not yet encounter regrouping. Have some interlocking cubes prepared in sticks of ten as well as some loose ones (page 21).

Warm-up
See Numberless word problems (page 18). Introduce this lesson with a numberless word problem, for example:

Some birds are sitting on a branch. Some of them fly away. Now there is a smaller number of birds.

- *What question can we ask about this story?*
- *What information would we need to solve the question?*

Gradually replace the numberless information with numeric information in order to build up to the question. For example, you can cross out 'some' in the first sentence and replace it with the number 29.

Ask a child to model this using cubes. Then ask: *How many birds are left? What information do we need to answer this question?* Once the children have worked out that we need to know how many flew away, replace the 'some' in the second sentence with the number 7.

Ask the child to model what we get if we take away 7 ones from the model of 29.

Invite the class to explain how we know how many are left.

Focus
You can use the same story frame with different numbers, or here are some more numberless story frames for subtraction:

- *There are lots of paintbrushes in a tin. The class takes some of the paintbrushes to use. Now there is a smaller number left in the tin.*
- *There are some sweets in a bowl. Some of the sweets are eaten. Now there are fewer sweets in the bowl.*

Here are some examples for addition:

- *There are some books on a shelf. The librarian puts more books on the shelf. Now there are lots of books on the shelf.*
- *A farmer picks some mangoes in the morning and places them in a crate. Later, she picks more and puts them into the crate with the others. Now there are more mangoes.*

As you work through the examples, you may need to remind the children of the signs we use for taking away or subtracting (−) and adding (+). It is important that they understand how to select which operation to use in order to solve a problem before deciding which sign represents that operation. In other words, start with the thinking and reasoning, not with the signs.

Once you have worked through several examples with the children with the concrete materials, work through questions 1 and 2 on **Pupil Book 2 page 36**. You can model these with cubes, or the children can count the cubes on the page.

For question 3, the children can work independently in pairs to work out the addition and subtraction sentences.

Support
- Display a large 100 chart in class and, if possible, prepare laminated or other wipe-clean versions for the children. Revise counting on in ones on the chart.
- Children could play board games such as 'Snakes and ladders' in which there is a board resembling a 100 chart. This will give them practice in simple addition and allow them to become familiar with the repeating pattern of numbers.

Answers for Pupil Book 2 page 36
1 a $45 - 4 = 41$ b $59 - 3 = 56$ c $34 - 4 = 30$
2 a $62 + 5 = 67$ b $34 + 4 = 38$ c $27 + 2 = 29$
3 a 84 b 78 c 41
d 59 e 31 f 29

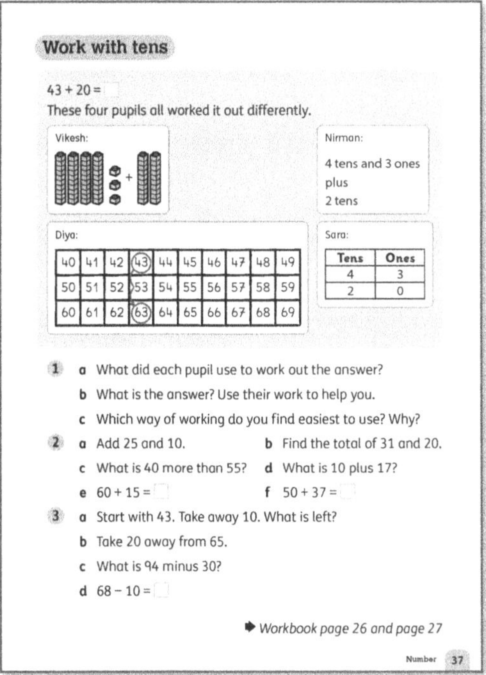

Materials
Base-ten blocks (page 22) or interlocking cubes
(page 21); 100 chart; place-value cards (page 22);
place-value tables (page 23)

Warm-up
- Give a word problem, for example: *I have 35 beans in
  a cup. I am going to add some more beans to the cup,
  in tens.*
- Model the number 35 using base-ten blocks or
  interlocking cubes sticks of 10 and single cubes. Ask:
  - *Can I work out how many beans I will have in
    total?* (No.)
  - *What do I need to know?* (You need to know how
    many tens to add.)
  - *OK, say I put in one more ten of beans. I had 3 tens
    and 5 ones and I'm adding one more ten. What
    have I got now?* (45)
  - *Let's take it away again and start again with 35.
    This time I want to add 2 tens to my 35. How can
    we work it out?*
- Repeat, adding different numbers of tens.

Focus
- Ask: *What other ways can we show this addition?*
  Ask whether anyone can show you on the
  100 chart. Then ask another child to show you using
  place-value cards. Also demonstrate adding using a
  place-value table:

Tens	Ones
3	5
2	0

- Continue working with examples in this way if
  needed. Then turn to **Pupil Book 2 page 37**. Talk
  about the four different ways that the children in
  question 1 worked out their answers.
- Work through question 2 and question 3 with the
  class. The questions are designed to develop the
  children's thinking around adding and subtracting
  tens and also their vocabulary. Draw their attention
  to the different ways we ask addition and subtraction
  questions:

  Addition: *find the total; what is ___ more than ___;
  what is ___ plus ___.*
  Subtraction: *what is ___ less than ___; find the
  difference between; take away.*

- Vary the placement of the multiple of 10 so that you
  are not always adding it second, and also vary the
  questions so that the larger number is not always
  placed first (for example, 12 + 50; 20 + 62).

Follow-up
Children can then work through **Workbook 2
page 26 and page 27** independently or with assistance, as
needed.

Support
Use the 100 chart. Revise counting on in ones and tens
on the chart, reminding the children that moving down a
row is 10 more and moving a across a column is 1 more.
Repeat for counting back.

Answers for Pupil Book 2 page 37
1 a Vikesh used sticks of 10 cubes and single
    cubes. Nirman decomposed/broke up the
    numbers into tens and ones. Diya used a
    number track or 100 chart. Sara used a place-
    value table.
  b 63
  c Individual's answers
2 a 35   b 51   c 95   d 27
  e 75   f 87
3 a 33   b 45   c 64   d 58

Answers for Workbook 2 page 26
1 a 39   b 43   c 65   d 79
  e 73   f 81   g 30   h 30

Answers for Workbook 2 page 27
1 a 26   b 27   c 21   d 17
  e 7   f 19   g 10   h 20

Work with tens and ones

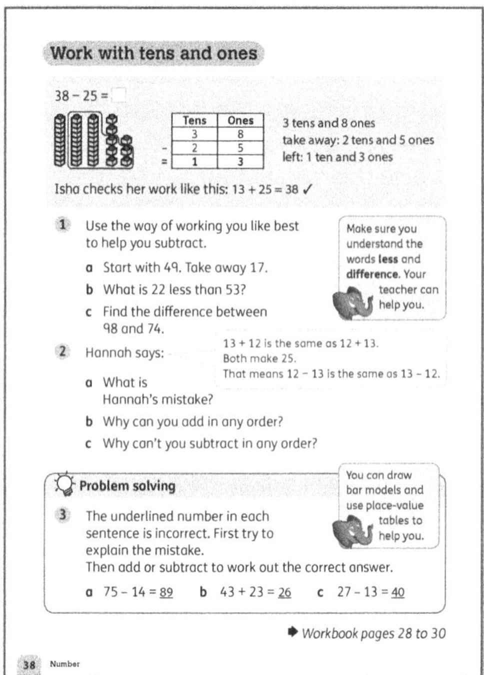

Materials
Base-ten blocks (page 22) or interlocking cubes
(page 21); 100 chart; place-value cards (pages 22–23)

In this lesson, continue working in the same way
as for the previous lesson, this time working with
combinations of tens and ones.

Warm-up
- Turn to **Pupil Book 2 page 38** and work through the
  example at the top of the page with the class.
- Work through some more addition and subtraction
  examples, always focusing on the wording and
  reasoning of the specific given problem, so that children
  get used to thinking about which operation to choose.

Focus
- Discuss the children's preferred ways of working to
  solve the subtractions in question 1.
- Discuss question 2 with the class. Reinforce the
  commutative nature of addition by asking the
  children to model 13 + 12 and 12 + 13 using
  base-ten blocks or cubes. Show how subtraction is
  not commutative in the same way.
- Problem solving: question 3 focuses on getting
  children to think about which operation they are using.

Follow-up
The children can work through **Workbook 2 pages 28–30**
to consolidate their work on addition and subtraction.

Challenge
Give some problems involving regrouping for children to
solve, for example:

$$55 + 19 \qquad 97 - 18 \qquad 50 - 12$$

Answers for Pupil Book 2 page 38
1 a 32   b 31   c 24

2 a She has changed the order of subtraction. You
can add in any order, but you can't take away
in any order.
    b The total of two numbers is the same,
whichever order they are added in. The order
doesn't matter.
    c When you are subtracting, the number left
depends on the number you take away. The
order matters.

<u>Problem solving:</u>

3 a mistake: added the numbers instead of
subtracting; correct answer: 61
    b mistake: added the ones correctly, but
subtracted the tens instead of adding them;
correct answer: 66
    c mistake: added the numbers instead of
subtracting; correct answer: 14

Answers for Workbook 2 page 28
1 a 59. Possible answer:

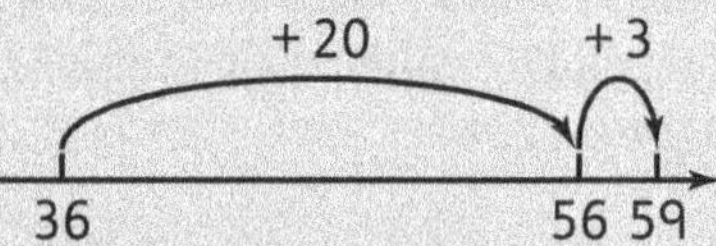

    b 42. Possible answer:

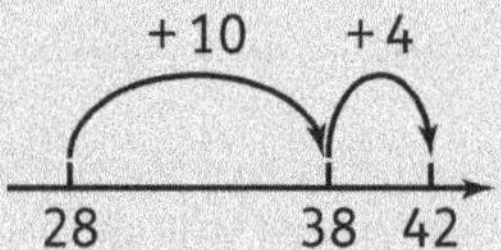

    c 52. Possible answer:

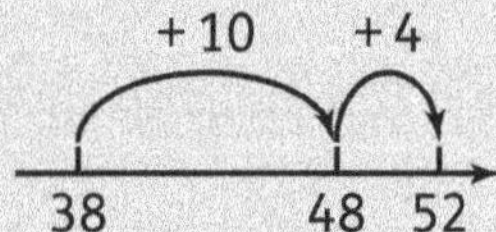

    d 50. Possible answer:

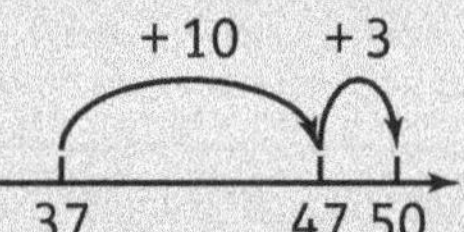

    e 63. Possible answer:

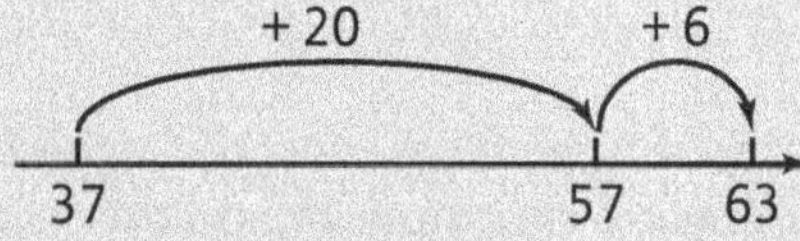

    f 72. Possible answer:

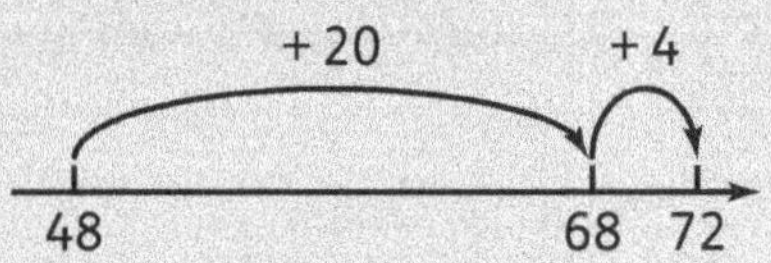

Answers for Workbook 2 page 29
1 a 19   b 15   c 66   d 11
    e 97   f 36

Answers for Workbook 2 page 30

1 **a** 10 and **90, 60** and 40, 70 and 30, 50 and **50, 80** and 20

  b 30 and **70, 40** and 60, 90 and **10, 20** and 80, 50 and 50

2 **a** 100 **b** 80 **c** 30 **d** 60
  e 50 **f** 100

3 **a** 90 **b** 10 **c** 80 **d** 20 **e** 70
  f 30 **g** 60 **h** 50 **i** 100

Add three numbers

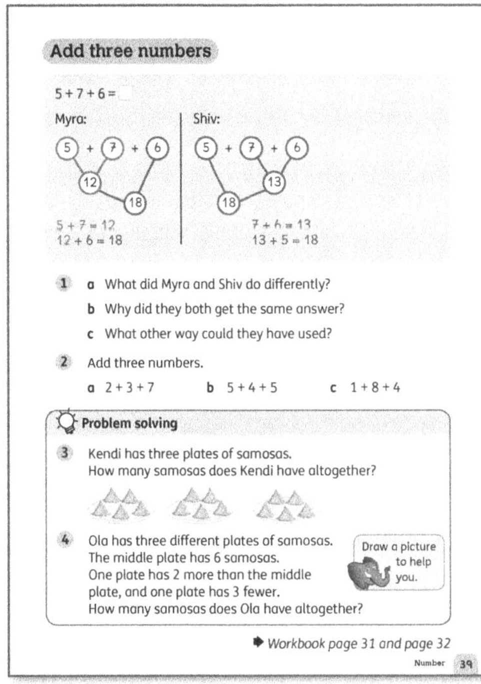

Materials

Stones or counters; lids or cups; chalk (for drawing a 100 grid outside – optional)

Warm-up

- Set out three lids or cups. Say: *I'm going to put some counters in each cup.* Ask children to volunteer a number under 20 for each cup and then count out that number. For example, put 8 counters in one cup, 9 in the next, 5 in the next. Ask: *How can I work out how many counters I have used altogether?*
- Let the children suggest ways to add them up. They may suggest putting all the counters together and counting them all. If so, agree that yes, this is one way of doing it. Ask whether anyone can suggest another way. Guide them to see that we can add together two of the numbers and then we can add another one.
- Once the children have done this (for example, 8 + 9 = 17, and 17 + 5 = 23), ask whether you could have added another way. Let them work out other ways of doing it (for example, starting with 8 + 5 or 9 + 5).

Focus

- If necessary, continue with more examples before working through questions 1 to 2 on **Pupil Book 2 page 39**.
- As children work through the questions, ask: *Which pair of numbers is easier to add together? Do you already know any number facts that can help you?* For example, for question 2 part a, they may already remember that 7 + 3 = 10, so that makes it easy to add the third number.
- Problem solving: The children can use concrete materials to model the samosas problems in questions 3 and 4, if they wish.

Follow-up

The children can work through **Workbook 2 page 31 and page 32**.

Challenge

Give the children more sets of three numbers to add, then sets of four or five numbers.

You can also make and say more problems similar to question 4, such as:

- *One bowl has 5 apples, one has 5 more than the first bowl and one has 5 fewer than the first bowl. How many apples are there altogether? (5 + 10 + 0 = 15)*
- *Jenny has three bunches of bananas. The first bunch has 6 bananas. The second has half as many. The third has double the bananas of the first. How many bananas are there altogether? (6 + 3 + 12 = 21)* (Note that this problem is extra challenging because the 'three bunches' and the ordinal numbers are extra number terms that the children have to ignore in their calculations.)

Support

- Draw a large 1–100 number grid on an area of the playground using chalk. Give groups of children instructions to move position on the grid corresponding to adding or taking away numbers.
- Revise counting on in ones and tens on the grid, reminding the children that moving down a row is 10 more and moving across a column is 1 more. Repeat for counting back.

Answers for Pupil Book 2 page 39

1 **a** They added the numbers in a different order. Myra started with 5 + 7 and Shiv started with 7 + 6.
  b The total of three numbers is the same, whatever order they are added in.
  c Start with 5 + 6, then add 7.

2 **a** 12 **b** 14 **c** 13

Problem solving:

3 15   **4** 8 + 6 + 3 = 17

Answers for Workbook 2 page 31

1 **a** $(3 + 7) + (9 + 1) = 10 + 10 = 20$

  b $(5 + 5) + (4 + 6) + 2 = 10 + 10 + 2 = 22$

  c $(8 + 2) + (1 + 9) = 10 + 10 = 20$

  d $(3 + 7) + (6 + 4) + 3 = 10 + 10 + 3 = 23$

  e $(6 + 4) + 3 + 5 = 10 + 8 = 18$

  f $(9 + 1) + (2 + 8) + 5 = 10 + 10 + 5 = 25$

  g $(9 + 1) + (6 + 4) + 8 = 10 + 10 + 8 = 28$

  h $(3 + 3 + 4) + (2 + 8) = 10 + 10 = 20$

  i $(8 + 2) + (7 + 3) + 7 = 10 + 10 + 7 = 27$

  j $(1 + 9) + (3 + 7) + 8 = 10 + 10 + 8 = 28$

Answers for Workbook 2 page 32

1 **a** 19 **b** 13 **c** 20 **d** 19

  e 19 **f** 30 **g** 23 **h** 20

  i 24 **j** 31

Round to tens

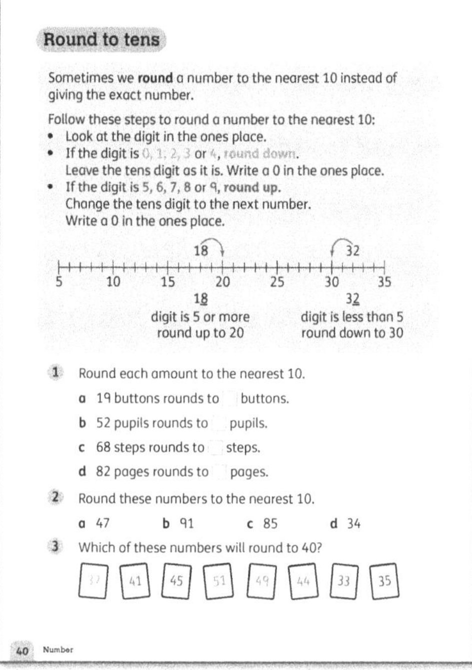

Materials
Beans; cups; place-value cards (pages 22–23); straws or sticks; bell or whistle

Warm-up
- Give each pair of children 20 beans and two cups. They put 10 beans in a cup to make a group of 10. One child says a number between 10 and 20. The other child uses the group of 10, together with other beans, to represent the number.
- Then one child uses beans to represent a number and the other child counts the number of beans to find the number.
- Repeat using place-value cards to relate the activity to tens and ones and place value.
- Extend the above activities using straws, sticks or beans to represent numbers between 1 and 50.

Focus
- Draw a large number line from 20 to 30 on the ground outside. Let pairs of children choose numbers and stand on them. When you ring a bell or blow a whistle, tell them to move to the 20 or 30 (whichever is closest to them). Have a discussion about what they should do if they are standing on 25. Make sure the children grasp that they have to move to the greater ten if they are on a multiple of 5.
- You may also wish to refer to use some of the 'Number work and operations' activities (page 28) on rounding numbers.
- Work through the information on **Pupil Book 2 page 40**. Explain the concept of *rounding* up and rounding down.
- Give children some practice in rounding different numbers before they complete questions 1–3.

Challenge
Give the children a round number and ask them to list all the numbers that will round to that number.

Interesting mistakes
Some children have difficulty interpreting the value of the digits in a 2-digit number. For example, if asked the value of the 4 in the number 47, some children will say it is 4. It will help their understanding if they think of a 2-digit number as the sum of so many tens and so many ones; that is, if they think of the number 47 as the sum of 40 + 7. Using place-value cards to build and partition numbers will help in this regard.

Answers for Pupil Book 2 page 40

1 **a** 20 buttons **b** 50 pupils

  c 70 steps **d** 80 pages

2 **a** 50 **b** 90 **c** 90 **d** 30

3 37, 41, 44, 35

Round to estimate

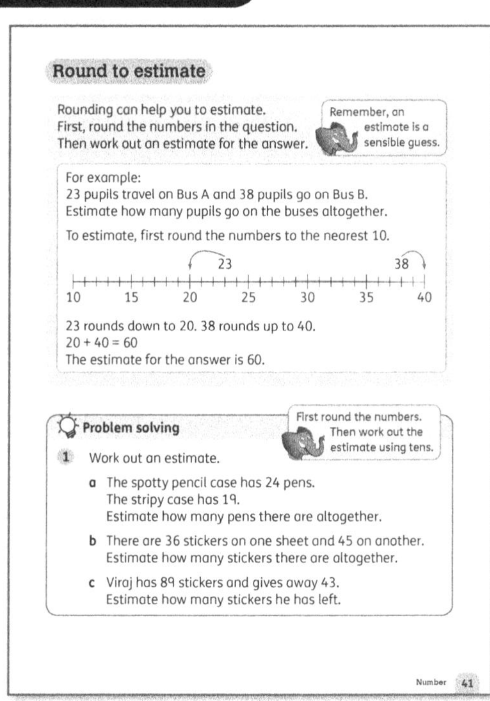

Materials

Prepare the materials needed for whichever activities you decide to use in the 'Warm-up'; chalk to draw a number line outside

Warm-up

Choose one or more 'Estimating' activities on pages 28 and 29.

Focus

- Work through **Pupil Book 2 page 41** with the class, going through the steps of how to estimate the answer to a problem by rounding each number first.
- Problem solving: Help the children apply the steps involved in estimating to complete each part of question 1.

Answers for Pupil Book 2 page 41

Problem solving:

1 **a** 20 + 20 = 40 (exact answer: 43)
   b 40 + 50 = 90 (exact answer: 81)
   c 90 – 40 = 50 (exact answer: 46)

End-of-unit check

Carry out activities to assess the children's understanding of bar models (see page 22):

- Show the children several bar models with one number missing. Ask the children to work out the missing number.

To assess the children's understanding of addition and subtraction:

Provide number sentences with a number missing, such as:

- 20 = 0 + ☐ (20)   4 + ☐ = 20 (16)   ☐ + 7 = 20 (13)
  20 = 5 + ☐ (15)

Ask these questions to assess the children's understanding of working with tens. (Replace the text in brackets with any suitable number.)

- *What is* (a multiple of 10) *plus* (a multiple of 10)*?*
- *What is 100 minus* (a multiple of 10)*?*
- *What is* (any 2-digit number) *plus* (a multiple of 10)*?*
- *What is* (any 2-digit number) *minus* (a multiple of 10)*?*
- *How can I quickly count on/back 40 on a number line?* (four skips of 10 either forwards or back)
- *How can I quickly add 40 to any number on the 100 chart?* (four skips of 10 forwards)

Ask questions to assess the children's understanding of place value, for example:

- *How many tens and how many ones are there in the number 29?* (2 tens and 9 ones)
- *A number is composed of 3 tens and 6 ones. What is the number?* (36)
- *What number is missing from the sequence 72, 73, ___, 75?* (74)
- *What symbol is used for add?* (+)
- *What is 23 take away 14?* (9)
- *What number appears immediately above 73 on a 100 chart?* (63)

Ask questions and carry out activities to assess the children's ability to add three numbers:

- *What is the total of* (any three numbers)*?*
- Ask the children to make sets of two, three or more numbers that they can add up to make a given total. They can use drawings, objects or counting apparatus to help them calculate.

Ask questions to move towards assessment of understanding of rounding and estimating:

- *How many ones are in 10?* (1)
- *How many ones are in 20?* (2)
- *What is the value of 2 in the number 32?* (2 ones)
- *What is the value of 5 in the number 57?* (5 tens)
- *Is 26 closer to 20 or to 30 (and so on for different numbers)?* (30)
- *True or false? The closest ten to 33 is 30.* (true)
- Give the children a range of 2-digit numbers and ask them to round each number to the nearest ten.
- Give additions and subtractions involving 2-digit numbers for the children to estimate the answers.

Learning objectives

- Solve problems with addition and subtraction using concrete objects and pictures, including those involving numbers, quantities and measures.

- Estimate and measure lengths of objects with non-standard or standard units.

- Choose and use appropriate standard units to estimate and measure length/height in any direction (m/cm).

- Compare and order lengths and record the results using >, < and =.

- Understand that length is a fixed distance between two points.

- Draw and measure lines using standard units.

- Understand that the markings on a ruler are a type of scale and connect them to a number line where intermediate points have value.

- Read from numbered scales on a ruler, starting at 0.

- Use the language of approximation.

Key words

length long wide high ruler metre stick standard units non-standard units centimetre metre estimate compare

Unit introduction

Materials
Some lengths of rope, string or ribbon; pens; paperclips

Teaching guidance
- Explain that *length* is how *long* something is. Hold up a length of rope, string or ribbon. Ask: *Is it long or short?* Lay it on a table. Then show another, and ask: *Is this one longer or shorter than the first?* Then hold up another. *What about this one? Is it longer or shorter?*
- Ask: *How can we measure how long something is? What does it mean to measure something?* Let the children give suggestions.
- Use a pen. Ask: *How many pens long is this piece of string?* Place the pen in the middle of the string. *Can I measure it like this? What is my mistake? Can someone show me how I work out how many pens long this piece of string is?* Encourage them to notice that we have to start by lining up at one end and then measure by carefully placing the pen at the point where it ended before.

- Ask the children to use paperclips to measure the lengths of short pieces of string and then arrange them in order from shortest to longest. Before they measure, ask: *About how many do you think it will be? What is your* estimate? Revise the term estimate, reminding them that they have seen it in the context of counting as well as addition and subtraction. Explain that we can also estimate measurements.

Talk about length

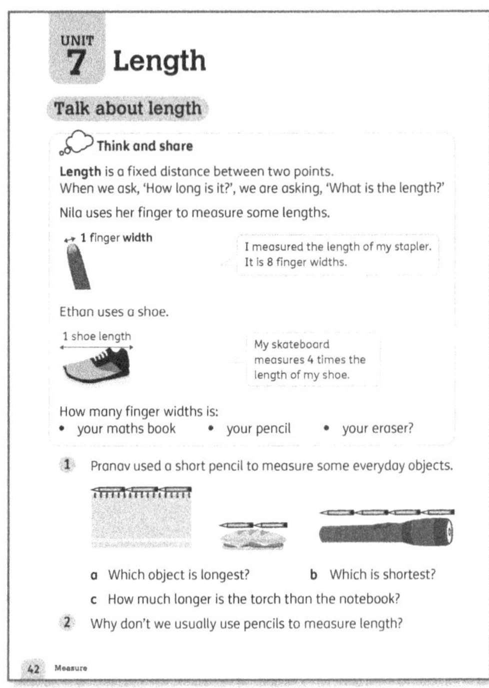

Materials
Small items to measure, such as books, pencils, boxes and shoes; small items to use as non-standard measures such as erasers and paperclips

Warm-up
Think and share: Discuss the example at the top of **Pupil Book 2 page 42**. Give the children some objects to measure in finger widths. You can use the examples suggested in the Pupil Book questions, as well as other objects around your classroom.

Focus
- Check the children's understanding of the term *compare*. They compare the lengths of the pictures in question 1 using a non-standard unit (pencils).
- For question 2, have a class discussion about the question: *Why do you think we don't usually use pencils to measure things?* Give real-life examples, using the terms *wide* and *high*, such as this one: *Imagine that a window breaks in my house. I need to get a new window, so I phone the glass repair company to order one. I say that the window is*

10 pencils wide/high. I am using this pencil (hold up a short pencil). *What do you think will happen if the person measures using a pencil at their office?*

Answers for Pupil Book 2 page 42

<u>Think and share:</u> Individual's answers

1 **a** torch   **b** sandwich   **c** 1 pencil

2 Different pencils are different lengths.

Centimetres

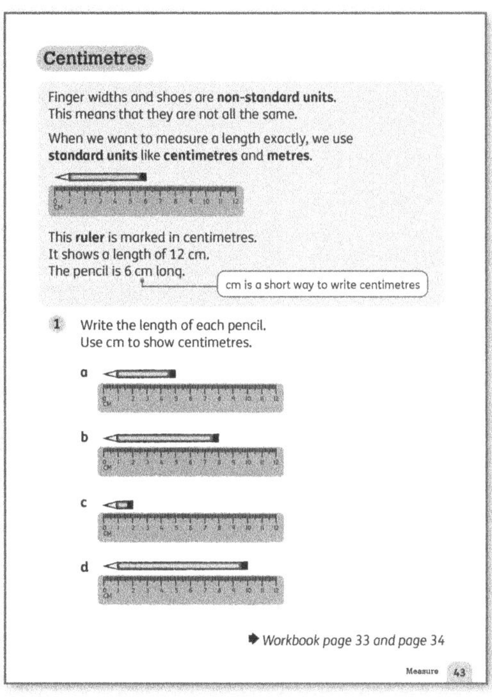

Materials

A set of four objects about a centimetre in length; base-ten blocks (10-rods and ones cubes only); objects for measuring such as lunchboxes, rulers and books; simple rulers marked in centimetres; curved objects for measuring; lengths of string; measuring tape (optional) 10 cm strips of paper marked in blocks for each child, as shown here:

1	2	3	4	5	6	7	8	9	10

Warm-up

- Use a set of four objects to show the children some objects that are about a centimetre in length. Allow them to find other examples and to check that these are about 1 cm long by comparing them to a ones cube or similar.

Focus

- Use a single ones cube to measure some small items in centimetres. Discuss the difficulty involved in moving the cube along to find the length and show that it is much easier to measure the length of an object with 10-rod, or paper marked in centimetres, because you can put it alongside the object and then count the number of units.

- Spend time estimating the lengths of different objects to reinforce the length of a centimetre and the length of 10 cm. Show an object and ask the children to say whether it is less than a centimetre, about a centimetre or more than a centimetre long. Do this for other lengths up to 10 cm.
- Introduce how to use a *ruler*. Ask the children measure small objects such as pens, books and lunchboxes. The idea is to familiarise them with the use of the ruler.
- When children look at a ruler, they need to understand that it is a type of scale. In other words, it uses a number line with markings between the whole units. Each of these intermediate points has a value. So, each of the small points between 0 cm and 1 cm is one of the 10 shorter lengths that make up one centimetre. At this stage, the children do not formally need to learn to use millimetres, but you can explain that centimetres are made up of these 10 smaller units.
- Try measuring some curved objects in centimetres. Try to get the children to suggest what the measurements will be, using strips of paper or lengths of string that they can then compare with a ruler. They may also suggest using a measuring tape, which is a practical solution.
- Go through the example on **Pupil Book 2 page 43** to introduce the terms *non-standard units* and *standard units*, giving the examples of *centimetres* and *metres* as standard units. Then ask the children to work through question 1.

Follow-up

Go through the instructions for **Workbook 2 pages 33–34** and make sure the children understand that they will need to gather up each of the objects listed and measure their width or length to the nearest centimetre using their finger, a ones cube, or a ruler.

Interesting mistakes

Make sure the children realise that they have to work from the 0 end of the ruler and count forward to get a measure; They cannot measure from the other end or start at an arbitrary point. Demonstrate measuring incorrectly and ask them to explain your mistake.

Answers for Pupil Book 2 page 43

1 **a** 5 cm   **b** 8 cm   **c** 2 cm   **d** 10 cm

Answers for Workbook 2 page 33

1 Individual's answer

Answers for Workbook 2 page 34

1 **a** about 15 cm   **b** about 100 cm
   c about 30 cm   **d** about 4 cm
   e about 15 cm   **f** about 30 cm
   g about 20 cm   **h** about 1 cm

2 Individual's answer

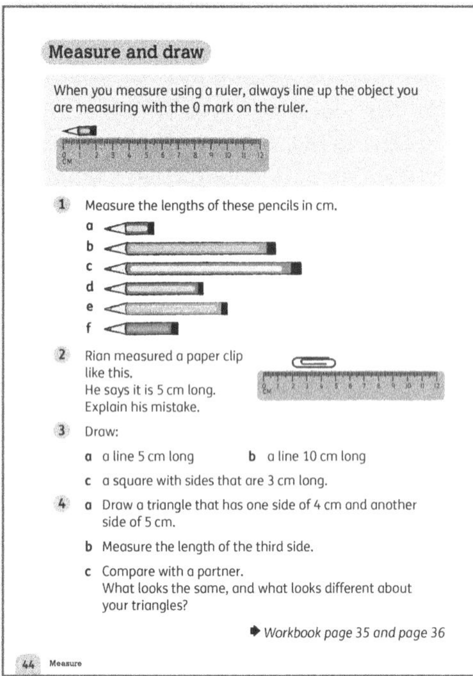

This page continues the work using rulers, but now gives children some practice in drawing lines of a given length.

Materials

Rulers

Warm-up

In order to draw lines, it is important to understand the need to line up the 0 on the ruler with the start point of the line. Demonstrate this on the board and give several children opportunities to draw lines of a given length using a ruler on the board.

Focus

Ask children to work through questions 1–4 on **Pupil Book 2 page 44** with a partner, then go through each question as a class.

- Ask the children to work out the total length of the sides of their triangles. They should record the total as a number sentence, including 'cm'.
- Ask the children to solve simple addition and subtraction problems, for example:
  ○ *I drew a line that was 14 cm long and then made it 8 cm longer. How long was my line?* (22 cm)
  ○ *Tia's ribbon was 24 m long and she cut off 6 cm. How long is her ribbon now?* (18 cm)

Follow-up

Workbook 2 page 35 and page 36 focus on measuring in centimetres. The children can work through these pages independently or with assistance.

Challenge

Give the children examples of 'measuring with a broken ruler', where you show segments of a ruler and objects placed at different start points (not 0). For example:

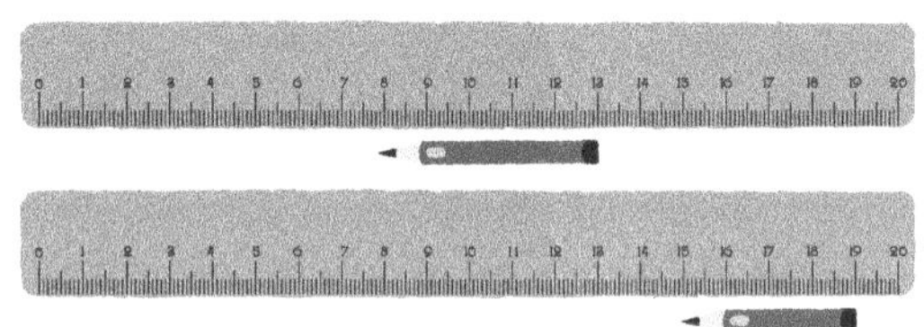

It is important that children understand that the 'broken ruler' activity will not give correct measurements if we just read off the numbers on the ruler. Instead, we have to work out how to count the number of centimetres as though the first centimetre was 0. The children may do this in different ways, for example as follows:

- They may treat the first calibration as 0 and then count on the number of centimetres.
- They may use subtraction to work out the number of centimetres of a given length.

Support

Give children more practice drawing lines of a given length, such as 12 cm, 15 cm, 10 cm. To help them, you can draw a start point and a straight line and let them use their ruler to mark where a line of a given length would end.

Answers for Pupil Book 2 page 44

1 **a** 2 cm   **b** 7 cm   **c** 8 cm
   d 4 cm   **e** 5 cm   **f** 3 cm

2 He has not lined up the start of the paperclip with the 0 mark on the ruler.

3 **a** a line 5 cm long   **b** a line 10 cm long
   c a square with sides of 3 cm

4 **a** and **b** Individual's answers
   c Both triangles should have one side of 4 cm and one side of 5 cm. The length of the third side may be different. The angles of the triangle may be different.

Answers for Workbook 2 page 35

1 **a** 3 cm   **b** 6 cm   **c** 12 cm
   d 1 cm   **e** 11 cm

2 **a** 5 cm   **b** 7 cm   **c** 10 cm
   d 2 cm   **e** 13 cm   **f** 4 cm

3 D, F, A, B, C, E

The metre

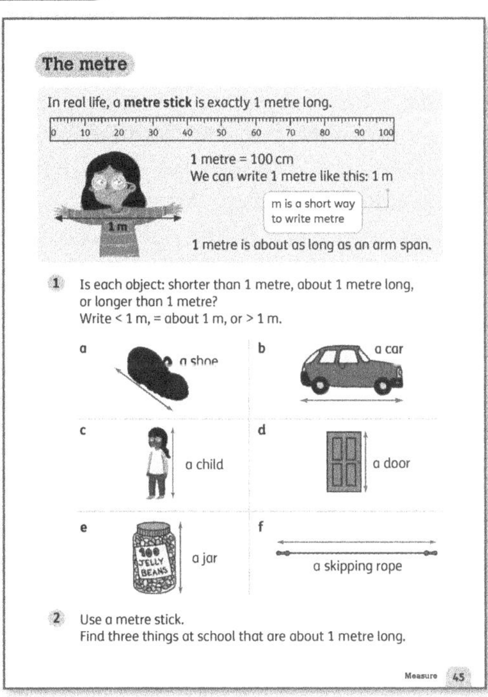

Materials

Metre sticks, rulers

Warm-up

Show the children a *metre stick*. Ask them to suggest what they think it is and what we might use it for. Ask:

- *For what things would this be useful to measure length?*
- *For what things would it be too long?*
- *For what things would it be too short?*

Say object names and ask children to decide if they should measure them in metres or centimetres. If possible, measure to check if they are unsure.

Focus

- If possible, organise children into small groups and give each group of children a metre stick and ask them to measure various lengths, heights and widths of objects around the classroom. They should estimate the length, width and height of the objects and then measure them to the nearest metre using a metre stick.
- Work through **Pupil Book 2 page 45** with the class. Revise the signs <, > and = with the children before they complete question 1.
- The children can work in their groups to complete question 2. Encourage them to estimate the length of the each thing first.

End-of-unit check

Ask questions to assess the children's understanding of length. For example:

- *About how many metres high is a door?* (2 m)
- *Mara sticks together a ribbon that 9 cm long and a ribbon that is 8 cm long. How long are they altogether?* (17 cm)
- *A piece of rope is 6 m long. I cut off 4 m. How long is the new piece?* (2 m)
- *What are the symbols for the metre and the centimetre?* (m and cm)
- *How many centimetres wide/long is your book?* (Answers depend on book chosen.)
- *Can you draw a line 9 cm long?* (line 9 cm long)

Learning objectives

- Choose and use appropriate standard units to estimate and measure mass (kg/g).

- Estimate and measure familiar objects using non-standard or standard units (kg/g).

- Begin to understand that mass is the quantity of matter in an object.

- Solve problems with addition and subtraction using concrete objects and pictures, including those involving numbers, quantities and measures.

- Compare and order mass and record the results using >, < and =.

Key words

lighter   heavier   equal   weight   mass   grams
kilogram   matter   quantity   balance scale

Unit introduction

Materials

Balance or shop scales (a simple balance can be made using a coat hanger and bulldog clips suspended on a hook or nail in a space where it can be free-moving); sand; bags large enough to hold 1 kg of sand; plastic 1-litre bottles; kitchen scale that weighs in kilograms; 2-kg bag of flour or rice; two bags; a variety of classroom objects to be weighed; 'weighs more/less' spinner

Teaching guidance

The correct scientific term for the weight of an object is *mass*. At this level though, we often use the term *weight*, as the children are physically weighing objects and it is the more commonly used term in everyday life.

- Spend some time revising the comparative language that was introduced in Level 1 (*heavy, light, bigger, smaller, lighter, heavier, lightest, heaviest*). You can do this by comparing objects and arranging them, or by allowing the children to lift objects and decide which is heavier/lighter. Include the terms *grams* and *kilogram*.

- Show the children two bags with some objects, one heavier than the other. Let them discuss which is heavier and which is lighter. Ask: *What makes something get heavier? How can I make this one heavier?* They may notice that if you put more things in the bag, it gets heavier. Ask why this is. *What makes something heavier?* Help them to notice that the more 'stuff' something has in it, the heavier it will be. They may also notice that a small object can sometimes make the bag heavier than a large object. Explain that *mass* is the term we use when we are measuring how heavy something is.

- Ask the children to look at the objects around them (desks, pens, pencils, their clothes, for example). Ask: *What is everything made from?* At first, they may identify the different materials things are made from. Encourage all suggestions. Then ask: *Does anyone know the word that scientists use for* all *the stuff that things are made from in the world?* Introduce the word *matter*. Explain that we say everything in the world is made from matter. Matter is anything that has mass, and mass is the word we use for the quantity of matter in an object. *Quantity* means how much.

- Before carrying out the practical work in this unit, the children must make their own 1-kg weights, using a plastic 1-litre bottle filled with water or sand.

- The children can make a simple balance by attaching a bag to each end of a coat hanger using bulldog clips and suspending the coat hanger from a hook or nail. A coat hanger balance must be suspended away from the wall so it is free-moving.

- The children should estimate the weights of a group of objects and then use 1-kg weights to check the accuracy of their estimates. Objects should be ordered in terms of increasing weight.

- Working in pairs, the children can use a 'weighs more/less' spinner. One child chooses an object and spins; the second child has to identify an object that weighs more or weighs less. This can be verified using a balance.

- The children should use a kitchen scale that is calibrated in kilograms to weigh some objects. Make sure they can read the scale, as they will need this skill in this unit.

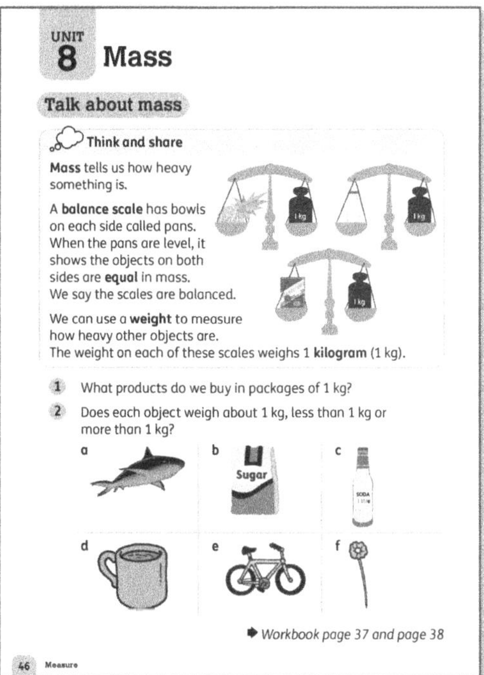

Materials

1-kg weight; balance scales; objects for weighing, including items that weigh about 1 kg (such as bags of beans, sugar, flour, fruits); apples (optional)

Warm-up

Let the children pick up the 1-kg weight. Ask them to guess which objects weigh more than/less than the weight. They can guess before and after picking up the objects. Then get them to check the weights using the balance scales.

Focus

- <u>Think and share</u>: Discuss the pictures at the top of **Pupil Book 2 page 46**, pointing out the explanation of a *balance scale*, and using the term *equal*. Ask the children to explain in their own words how the scales work. They should understand that the pans are level when both sides are equal, so in these pictures each item is equal to 1 kg.
- For question 1, give the children the opportunity to hold the items you have brought in that weigh about 1 kg. Have a class discussion about products that might come in packages of about 1 kg.
- Talk with the children about the pictures in question 2. Ask them to estimate whether these items would weigh about 1 kg, less than 1 kg or more than 1 kg.

Follow-up

- Ask the children to look at **Workbook 2 page 37**. Encourage them to draw the things that they estimate are heavier or lighter before getting them to use a balance to check. Provide apples for weighing if possible.

- The children can work independently to read the scales and record the amounts in kilograms on **Workbook 2 page 38**. Allow them to use the abbreviation kg if they want to.

Support

For children who do not yet have a clear idea of how much a kilogram weighs, give them plenty of practice picking up various things that weigh 1 kg. Let them choose items that are lighter, and put those into a group, and then items that are heavier, and put those into a group. At the stage, the children do not need to have an accurate sense of mass. They just need to have an idea of how much is 'about 1 kg'.

Answers for Pupil Book 2 page 46

1 Possible answers: a bottle of water, a bag of rice, etc.

2 a more than 1 kg

  b either less than 1 kg, about 1 kg or more than 1 kg

  c about 1 kg   d less than 1 kg

  e more than 1 kg   f less than 1 kg

Answers for Workbook 2 page 37

1 Individual's answers

Answers for Workbook 2 page 38

1 a 3 kg   b 2 kg   c 1 kg

  d 5 kg   e 3 kg   f 4 kg

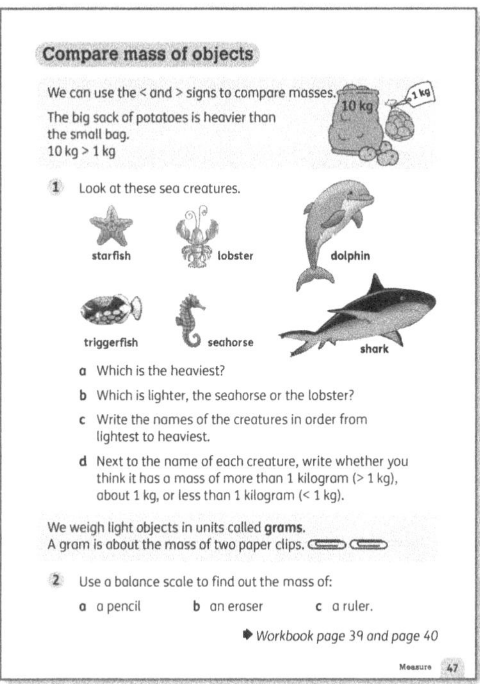

Materials

1 kg weight; balance scales; a variety of objects that weigh between 500 g and 2 kg (such as different fruit and vegetables); a kitchen scale (preferably an electronic one); craft materials (e.g. sticks, cardboard, tape – optional)

Warm-up

Show the objects to the children and let them guess which items weigh more than, less than or about 1 kg. They can sort the items into these three sets. Next, let them check by picking up and judging by feel. They can move items to a different set if they change their minds. Lastly, check as a class by comparing the objects using the balance scales.

Focus

- Revise the < and > signs. Say that some children think that the less-than sign (<) looks a bit like an L, which can help to remember that it is the less than sign.
- Talk about the example at the top of **Pupil Book 2 page 47** before discussing the sea creatures in question 1. Ask the children to tell you what they know about each of them. If the children are not familiar with all the creatures in the picture, you may want to show them some material online to help them get an idea of their relative sizes and masses.
- The children move on to practical weighing activities in grams in question 2.

Follow-up

Workbook 2 page 39 and page 40 consolidate the learning from the Pupil Book pages. For **page 39**, let the children work in pairs to do the actual measuring. **Page 40** is more challenging only because the masses extend to 100 kg. Allow the children to work in pairs to solve the problems and complete the page.

Challenge

Give the children some craft materials and ask them to make:

- the smallest object they can make that has a mass of 1 kg
- the largest object they can make that has a mass of 1 kg.

Answers for Pupil Book 2 page 47

1 **a** Possible answer: shark
   b Possible answer: seahorse
   c Possible answer: seahorse, starfish, triggerfish, lobster, dolphin, shark
   d seahorse < 1 kg, starfish < 1 kg, triggerfish < 1 kg, lobster about 1 kg, dolphin > 1 kg, shark > 1 kg
2 Individual's answer

Answers for Workbook 2 page 39

1 Individual's answers
   a and **d** The children should draw objects that have a mass of 1 kg.
   b and **f** The children should draw objects that have a mass greater than 1 kg.
   c and **e** The children should draw objects that have a mass less than 1 kg.

Answers for Workbook 2 page 40

1 jaguar 90 kg, leopard 65 kg, cheetah 45 kg, coyote 20 kg, lynx 15 kg, house cat 5 kg
2 **a** heavier   **b** leopard
   c house cat   **d** lynx
3 Possible answers:
   a elephant   **b** mouse

Solve problems

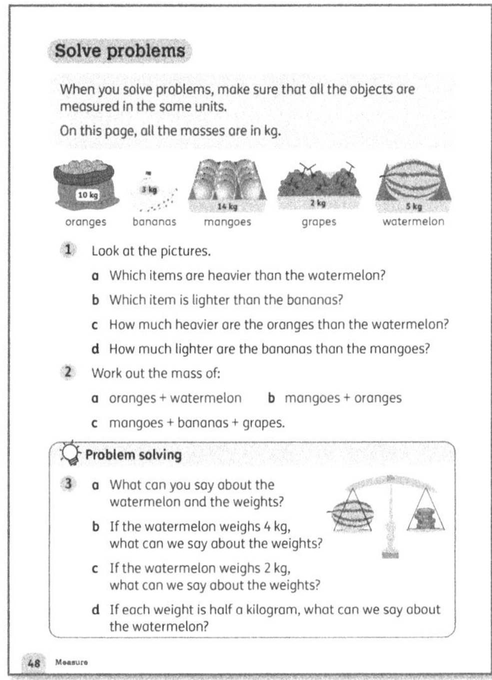

Materials

Counters

Focus

- You may choose to work through all the problems on the page with the class, or work through question 1 and question 2 part a, before letting the children complete part b and part c on their own.
- For question 1 part c and part d, you may need to explore with the class how to work out how much heavier or lighter one thing is than another. Do not be too hasty in offering subtraction as the way of solving this. Instead, ask an open-ended question: *How can we work this out?*

- You may want to give the following simple example first and present it as a number talk (see Number talks, pages 17–18). *This object weighs 2 kg. This object weighs 3 kg. How much more does the second object weigh than the first? How do you work it out?*
- Suggestions could include counting back along a number line and subtracting to find the difference.
- <u>Problem solving:</u> For question 3, the children need to discuss which is heavier – the melon or the weights – and what this could tell us. We don't have any exact weights, but we can make statements using 'more than' and 'less than'.

Challenge

Set some additional problems, such as these:

- Ask the children to look at the picture of the mangoes on **Pupil Book 2 page 48**. Say: *This picture shows 12 mangoes. The tray of mangoes weighs 14 kg. About how much do you think each mango weighs?* (Just over 1 kg)
- Ask the children to use the masses of the animals on **Workbook 2 page 40**. Give them these instructions: *Draw a number line. Label the left end heaviest and the right end lightest. Mark points on the line to position the masses of the animals correctly.*

Interesting mistakes

Some children may make the mistake of thinking that asking how much heavier or how much lighter determines the operation as addition or subtraction. Give more simple examples to demonstrate the concept of difference: For example, say: *A melon weighs 5 kg and a pumpkin weighs 7 kg. Show this using, say 5 counters and 7 counters. Ask:*

- *How many more counters are on this side?* (2)
- *How many counters fewer are on this side?* (2)
- *What is the difference between 5 and 7?* (2)
- *How much more does the pumpkin weigh?* (2 kg) *How much less does the melon weigh than the pumpkin?* (2 kg) *Why do both questions give the same answer?* (The difference between the melon and pumpkin is 2 kg. The order does not matter.)

Answers for Pupil Book 2 page 48

1 a oranges, mangoes   b grapes
   c 5 kg   d 11 kg
2 a 15 kg   b 24 kg   c 19 kg

<u>Problem solving:</u>
3 a They have the same mass.
   b Each weight is 1 kg.
   c Each weight is $\frac{1}{2}$ kg.
   d It weighs 2 kg.

Ask questions to assess the children's understanding of mass, for example:

- *What does kg stand for?* (kilogram)
- *Can you arrange these weights in increasing order of mass?* (Give the children five different weights.)
- *How do you know which object is heavier on a balance scale?* (The heaviest item will make that side of the scale lower.)
- *A box of bricks has a mass of 12 kg. A box of tiles has a mass of 9 kg. Which is heavier?* (The box of bricks) *How much heavier?* (3 kg)
- *A small packet has a mass of 4 kg. A larger packet has a mass of 7 kg. What is their mass altogether?* (11 kg)
- *Jossie weighs 10 kg more than Mika. If Mika weighs 23 kg, how much does Jossie weigh?* (33 kg)

Mixed practice 1

Answers to Mixed practice 1 on Pupil Book 2 pages 49–50

1 Possible answers: a row of 6 counters; a row of 5 counters + 1 counter; a row of 4 counters + a row of 2 counters
2 a second: tomatoes   third: pears
     fourth: pineapples
   b coconut milk: fifth   corn: ninth
   c Individual's answers
3 a forty-eight; 4 tens 8 ones
   b ninety-two; 9 tens 2 ones
   c eighty; 8 tens 0 ones
   d thirty-nine; 3 tens 9 ones
   e twenty-five; 2 tens 5 ones
4 a 4, 11, 18, 23   b 23, 32, 35, 53
5 a circle   b rectangle
6 a cube   b sphere   c cuboid   d pyramid
   e cylinder   f cone
7 Individual's answers. The explanation should use the letters A, B, C, for example: ABC, ABBC (to match the sequence of the three shapes).
8 5, 10, 15, 20, 25, 30, 35, 40, 45, 50
9 a 5 cm > 3 cm   b 2 cm < 6 cm
   c 1 cm < 2 cm
10 25 cm
11 25 cm
12 a The height of Kevin's tower
   b 12 cm
13 a less than 1 kg
   b equal to 1 kg
   c more than 1 kg

Learning objectives

- Interpret and construct simple tally charts.

- Ask and answer simple questions by counting the number of objects in each category and sorting the categories by quantity.

- Ask and answer questions about totalling and comparing categorical data.

- Ask and answer simple questions by counting the number of objects in each category.

Key words

data   table   row   column   tally mark
tally table   Venn diagram   Carroll diagram
categories   even   odd

Unit introduction

Some curriculums require children to use information presented in simple *tables*. Children do this throughout this course, as information is frequently presented in table form. Make sure they are familiar with the concepts of *rows* and *columns*, as well as headings for tables.

Teaching guidance

- Ask the children to think of as many different animals as they can. Let them suggest 15 or so different kinds of animal. Encourage them to think of some animals that can swim, some that can fly, some that can run and animals that jump or hop. Add some of your own suggestions, such as duck, frog, rabbit, pigeon.
- Then draw a circle and label it 'Animals that fly'. Ask the children to list the animals that can fly. Write the names of those animals in the circle. You can also include insects.

Explain that we call this a *set* and animals that fly fit into this set.

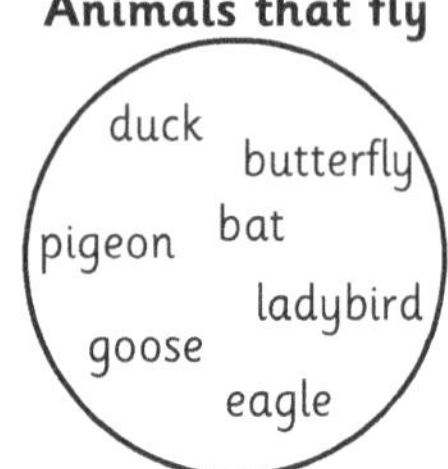

- Draw another circle and label it 'Animals that swim'. Ask which animals fit into this set. Invite them to identify animals that can swim. Write the names in the circle.

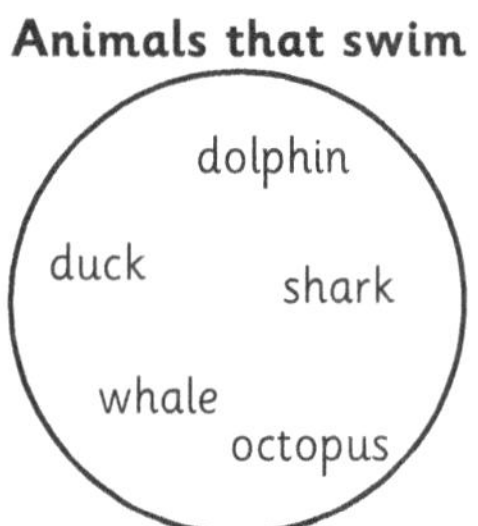

- Then ask: *Are there any animals that fit into both groups?* (For example, a duck can fly and it can swim.)
- Ask: *Can anyone think of a way we can show these sets in a way that shows which animals fit into each group and which fit into both?* If any children have an idea how to do this, let them come up to the board and show the rest of the class. They may already have an idea of how to do an overlapping Venn diagram, or they may show it in a different way. Discuss the ideas with the class.
- Show children how to draw two overlapping circles so that elements that fit into both sets go in the overlapping section (the intersect), as follows.

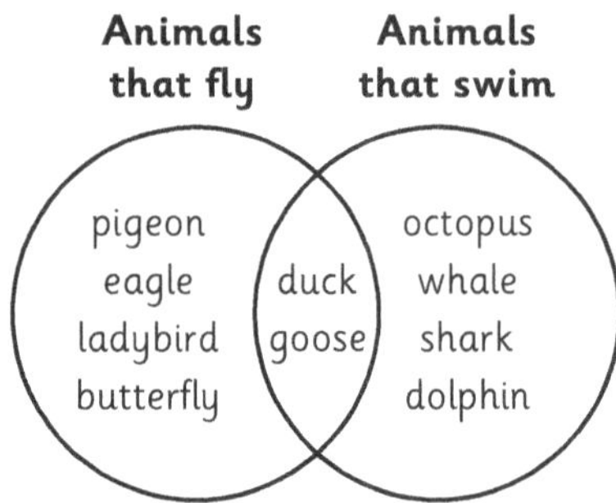

- Then ask: *What other ways could we sort the animals in this list?* Encourage other suggestions, such as animals that lay eggs; big and small animals.

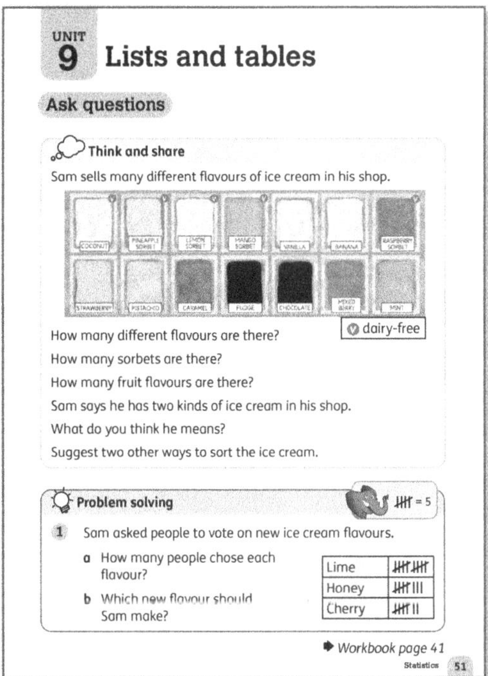

Warm-up

- Think and share: Turn to **Pupil Book 2 page 51** and ask the children to describe what they can see in the picture. Work through the questions in a class discussion. Ask the children to estimate the number of flavours before counting. The rest of the questions focus on the different ways of sorting the flavours.

You may need to explain that a sorbet is an ice cream that is made without milk or cream, just with a sugar syrup, so it is more like an ice lolly.

Focus

- Work through how to draw *tally marks*. Explain that each mark represents 1 but that we show 5 by drawing the fifth mark as a strike through the first 4 marks:
  | = 1 ||||| = 5
- Draw several bundles of 5 and ask the children to revisit counting in fives.
- Problem solving: Explain what *tally marks* and *tally tables* are, referring to the example tally table in question 1. The children use this tally table to answer the question.

Follow-up

On **Workbook 2 page 41**, the children need to count how many there are of each fruit or vegetable, and fill in the table. They use the middle column for ticks and the last column for the totals.

Answers for Pupil Book 2 page 51

Think and share: There are 14 different flavours of ice cream.

There are 4 sorbets: pineapple, lemon, mango, raspberry.

There are 8 fruit flavours: coconut, pineapple, lemon, mango, banana, raspberry, strawberry, mixed berry. (Coconut is a type of fruit, not a nut as its name suggests.)

Possible answer: Sam has sorbets and ice creams in his shop.

Possible answer: fruit flavour and not fruit flavour ice creams

Problem solving:

1 a   lime: 10; honey: 8; cherry: 7   b lime

Answers for Workbook 2 page 41

1 Apples 9; bananas 13; oranges 10; carrots 15; onions 9; potatoes 17. Total: 73

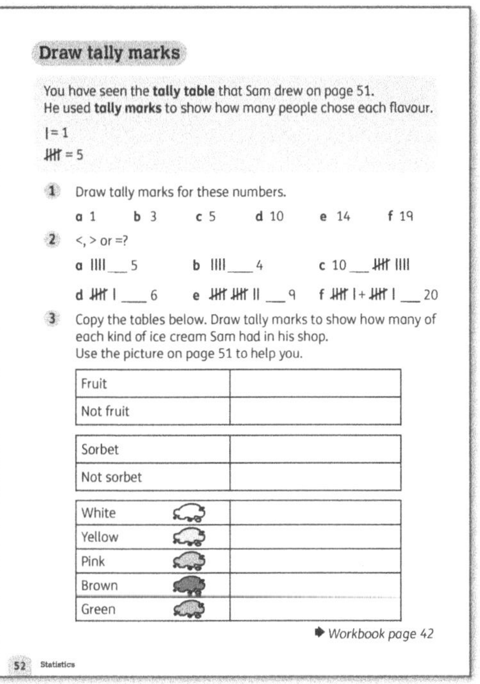

Warm-up

- Practise counting forwards and backwards in fives.
- Say a number that is not a multiple of 5, for example, 17. Ask the children to count up to 17 in fives and then ones. (five, ten, fifteen, sixteen, seventeen) Repeat with other numbers.

Focus

- Draw tally marks for different numbers between 1 and 30 and ask the children to count up what each tally represents.
- Give individual children opportunities to draw the tally marks for given numbers. Focus on the need to work out 'how many fives' and 'how many more' for each number.
- Work through questions 1–3 on **Pupil Book 2 page 52** with the class.

Follow-up

Workbook 2 page 42 gives extra practice with tally tables.

Answers for Pupil Book 2 page 52

1   a | b ||| c   ЖЖ   d   ЖЖ ЖЖ
    e   ЖЖ ЖЖ |||| f   ЖЖ ЖЖ ЖЖ ||||

2   a   <   b   =   c   >
    d   =   e   >   f   <

3   1st table: fruit: ЖЖ ||| not fruit: ЖЖ ||
    2nd table: sorbet: |||| not sorbet: ЖЖ ЖЖ
    3rd table [Possible answer]: white: |||| yellow: ||
    pink: ||| brown: ||| green: ||

Answers for Workbook 2 page 42

1   triangle ЖЖ ЖЖ ||, 12
    circle ЖЖ ЖЖ |||, 13
    square ЖЖ ЖЖ ЖЖ, 15

2   Individual's answers

Venn diagrams

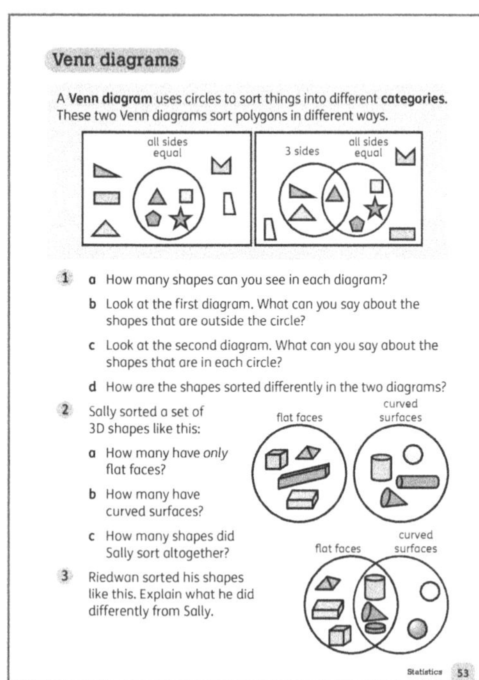

Materials

Objects for sorting into Venn diagrams; large rings or print-outs of overlapping circles

Warm-up

- Do some practical sorting tasks with the class to revise the concept of a *Venn diagram*. Objects can be sorted in many ways: by shape and colour; shape and size; yellow and red fruits; leaves with straight and curved edges; wheels and no wheels; and so on.

Focus

- The children may not be familiar with the word *categories*, although they will have sorted things into different categories. Explain that categories is just another word for the groups that we arrange things into, such as different shapes, colours and sizes.
- Use questions 1–3 on **Pupil Book 2 page 53** to teach the main concepts:
  - Venn diagrams use circles to show sets.
  - Everything that belongs in the set goes inside the circle. Everything that does not belong goes outside the circle.
  - Two circles can overlap to show that some objects fit into both sets.

Challenge

The children can create their own overlapping Venn diagrams for sorting items such as stationery, food items or clothing.

Support

Some children may need extra practice in sorting objects into Venn diagrams. First, allow them to practise sorting things into a defined set (one circle). Here are several examples you could use:

- Provide a variety of objects of different colours, and ask them to show the set of objects of a given colour, for example red objects.
- Recall the activities on mass from Unit 8, and draw sets of different objects that weigh more than 1 kg/about 1 kg/less than 1 kg.
- Sort coins into sets of silver and copper coins.
- Sort wooden/plastic 2D shapes into sets of triangles, circles and rectangles.

Once they have had more practice sorting items into defined sets, you can draw overlapping circles and let them sort according to two categories. You can use any ideas that fit with the themes you are currently using in class. The following are suggestions only:

- triangles; black shapes
- fruits; things that weigh about 1 kg
- things we can drink; foods that contain sugar
- fruits; yellow things

Answers for Pupil Book 2 page 53

1 a 9

   b They are polygons that do not have all sides equal.

   c In the first circle, they are polygons with three sides. In the second circle, they are polygons with all sides equal. The triangle in the middle (intersection) has three sides and all sides equal.

   d In the first diagram, there is only one category. In the second diagram, there are two categories, which overlap.

2 a 4   b 4   c 8

3 He sorted the same eight shapes, but he showed which shapes have both flat faces and curved surfaces (in the intersection).

Carroll diagrams

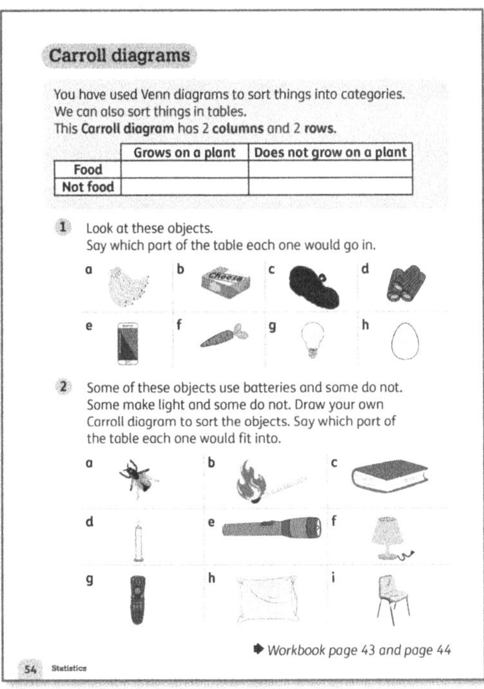

Materials

Items for sorting such as plastic shapes, large and small objects, stones, sticks, natural and human-made objects

The children have already started looking at using two criteria in Venn diagrams. Using two sets of criteria is the concept used to develop *Carroll diagrams*. A Carroll diagram is a type of two-way table that relies on the idea of objects being either one thing or not that same thing, to sort information. For example, shapes could be classified as square or not square as well as blue or not blue. Objects fit into boxes on the table if they meet both criteria specified for a box.

Warm-up

Spend some time physically sorting and classifying objects using the criterion of being or not being a particular thing. For example, you could sort items into 'plastic' and 'not plastic'. It is important the children focus on the 'plastic' criterion and that they realise that all objects that are not plastic fit into the 'not plastic' group, no matter what they are made of.

Focus

- Use the Carroll diagrams in question 1 on **Workbook 2 page 43** to teach the children how to combine two Carroll diagrams to get one Carroll diagram that compares the items using both the criteria used in the first two Carroll diagrams. Before they start, check that they understand the terms *even* and *not even* (*odd*). In the third Carroll diagram, make sure they realise for example that in order to fit into the top-left box, the number has to be both even *and* less than 10.
- Then work through the example and questions 1 and 2 on **Pupil Book 2 page 54** with the class.

Follow-up

Use **Workbook 2 page 44** to make sure the children can distinguish shapes based on two attributes (colour and shape) and to assess whether they are able to use a Carroll diagram to sort data. You might want to work through the page with the class or let them complete it independently.

Answers for Pupil Book 2 page 54

1

	Grows on a plant	Does not grow on a plant
Food	a banana f carrot	b cheese h egg
Not food	d wood	c shoe e smartphone g lightbulb

2

	Uses batteries	Does not use batteries
Makes light	e torch	a firefly b match d candle f lamp
Does not make light	g TV remote control	c book h pillow i chair

Answers for Workbook 2 page 43

1

Even	Not even
12, 8, 18, 20	19, 5, 11, 7, 3, 17, 15, 1

2

Less than 10	Not less than 10
5, 8, 7, 3, 1	19, 11, 12, 18, 20, 17, 15

3

	Even	Not even
Less than 10	8	5, 7, 3, 1
Not less than 10	12, 18, 20	19, 11, 17, 15

Answers for Workbook 2 page 44

1

	Grey	Not grey
Square	▢ ▫ ▢	▫
Not a square	◯ ▮	▽△◯ ▭

Sort data

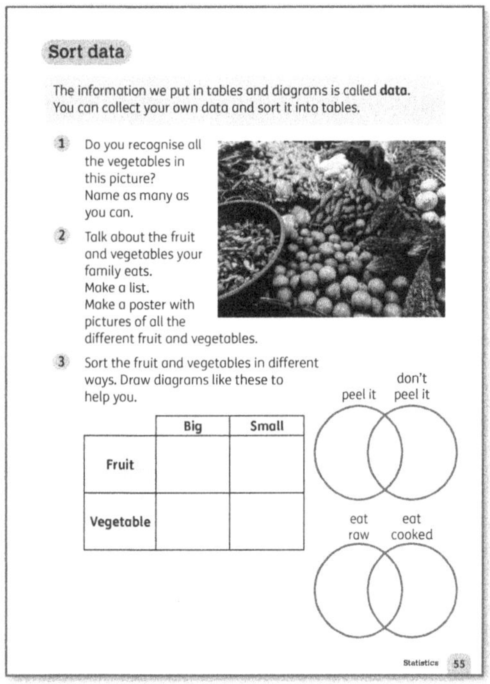

Warm-up

- Start the lesson by pointing out that any information we put into tables or diagrams is called *data*. Complete question 1 and question 2 on **Pupil Book 2 page 55** with the class.
- Discuss the vegetables in the picture with the children, as well as the vegetables they eat at home. Write a class list of all the vegetables on the board.

Focus

- For question 3, help the children to draw a variety of tables and diagrams to sort vegetables.

Answers for Pupil Book 2 page 55

1 chillies, peppers, carrots, banana blossoms, beans, limes, karela, daikon radishes

2 and **3** Individual's answer

End-of-unit check

Give the children activities to assess their understanding of tables and diagrams, for example:

- Give the children numbers to write as tally marks. (For example, 3 is ||| and 6 is ⊣⊦⊦ |).
- Give tally marks for the children to interpret as numbers. (For example, ⊣⊦⊦ is 5 and ⊣⊦⊦ ⊣⊦⊦ | is 11).
- Give the children simple blank Venn diagrams and Carroll diagrams, like the ones below and ask them to write or draw items that fit into each set:

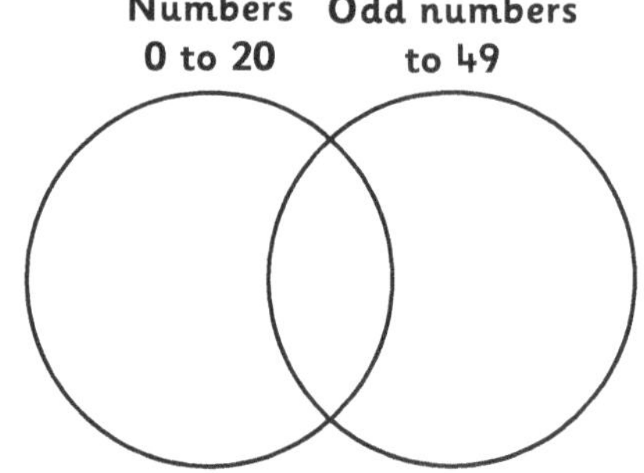

	Found in water	Not found in water
Animal		
Not an animal		

Learning objectives

- Intrepret and construct simple pictograms, tally charts, block diagrams and simple tables.
- Ask and answer simple questions by counting the number of objects in each category and sorting the categories by quantity.
- Ask and answer questions about totalling and comparing categorical data.

Key words

chart   pictogram   block diagram
key   row   column

Unit introduction

Materials

Pictogram showing favourite days of the week, as shown below; large grid for block diagram, as shown below

Monday	👤 👤 👤 👤 👤 👤
Tuesday	👤 👤
Wednesday	👤 👤 👤 👤
Thursday	👤 👤 👤 👤 👤 👤
Friday	👤 👤 👤 👤 👤 👤 👤 👤 👤 👤 👤 👤
Saturday	👤 👤 👤 👤 👤 👤 👤
Sunday	👤 👤 👤 👤 👤

Key: 👤 = 1 student

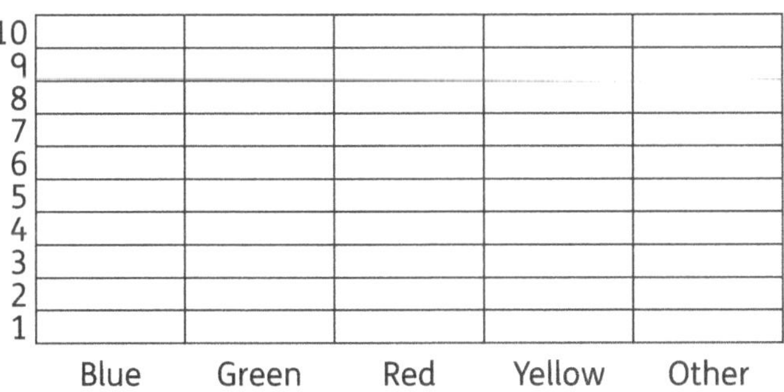

Favourite colours

Block diagram grid with vertical axis numbered 1 to 10 and horizontal axis labelled: Blue, Green, Red, Yellow, Other.

Teaching guidance

- Revise the concept of a *pictogram* using the 'Favourite days of the week' pictogram. Ask questions such as: *What can you see in the first* column? *What do we know when we look at the second row? Which day did most children like? Which day did the fewest children like best? How many children liked Monday? How many more children liked Tuesdays than Wednesdays?*
- Make a horizontal display on the board. Write the letters of the alphabet in a horizontal row across the top of the board. Get the children to put up their hands to show what letters their first names begin with. Draw a stick figure below the relevant letters to represent each child. Do this systematically; for example, say: *Everyone whose name starts with A, put up your hand.* As you draw each child, ask them to sit down. Use the words *pictogram* (a type of *chart*) and information in context to reinforce their meaning. Ask the children questions about the pictogram.
- Revise tally tables. Draw this table on the board:

Favourite colour	Number of children
blue	
green	
red	
yellow	
other	

- Ask each child to choose their favourite colour from the list and record each vote with a tally mark in the table. Discuss why there is an 'other' row (because it is not possible to list every colour in a table).
- Revise the use of the words *row* and *column* if necessary.
- Revise *block diagrams*, by drawing a block diagram of the favourite colour data on a large grid.
- Draw columns of blocks to represent the number of children who chose each colour. Then rotate the chart through 90° so the information is shown by rows of blocks. Explain to the children that you can use blocks in columns or in rows in a block diagram. Keep this block diagram for the lesson on 'Block diagrams' (page 90).
- Make a permanent weather chart for the classroom. It could look something like this one:

	Monday	Tuesday	Wednesday	Thursday	Friday	Saturday	Sunday
Today it is		☀					

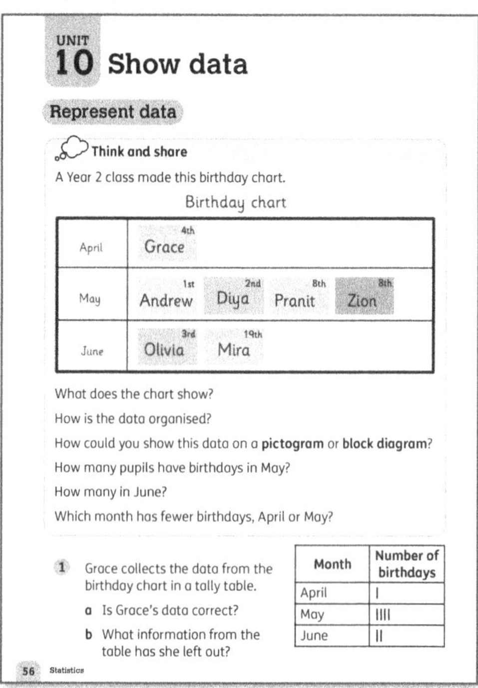

Warm-up

<u>Think and share:</u> Turn to **Pupil Book 2 page 56** and discuss the picture of the birthday chart with the class. Ask the children to describe what information they see in the chart. If necessary, guide the discussion with questions such as:

- *How many rows are there?* (3)
- *How many columns are there?* (2)
- *What do you see at the beginning of each row?* (months)
- *What do you think the numbers on each sticky note tell us?* (Pupils's names and birthday dates.)

Focus

Work through question 1 with the class.

Answers for Pupil Book 2 page 56

<u>Think and share:</u> The chart shows which pupils in the Year 2 class have their birthdays in the months April, May and June.

The data is organised in rows, with one month in each row. The name of the month is in the first column and the pupils' names and birthday dates are in the next column. The names are organised in date order.

A block diagram or pictogram will have the same layout. Instead of the names, a block diagram will have one block for each birthday; a pictogram will use pictures and will have a key to say how many birthdays each picture represents. Possible answer:

Month	Number of birthdays
April	😊
May	😊 😊 😊 😊
June	😊 😊

Key: 😊 = 1 birthday

Four pupils have birthdays in May.
Two pupils have birthdays in June.
April has fewer birthdays than May.

1 a yes
b the dates of the birthdays

Pictograms

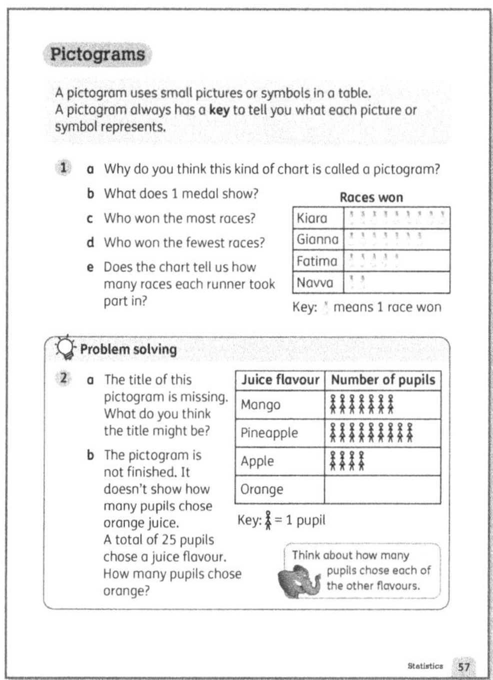

Materials

A shape pictogram worksheet (as shown here) for each group of children; sets of shapes

Circle						
Square						
Rectangle						
Triangle						

Warm-up

- Give each group a set of shapes and a shape pictogram worksheet. Agree on a symbol to use to represent all the shapes, for example a smiley face. They can draw symbols to represent the shapes in their set. Remind them to make a *key* to show what one symbol represents.

> Some children may ask whether a square is also a rectangle. In mathematics, a *rectangle* is a quadrilateral with four right angles and opposite sides that are equal in length. As a square conforms to this definition, it is technically also a rectangle. A *square* is a 'special' rectangle that has all four sides equal in length. If the children do not raise this question, you may want to leave the topic for now.

Focus

- Work through question 1 on **Pupil Book 2 page 57** with the class.
- You may also want to discuss how the Races Won pictogram is similar and different from the pictograms they made in the Warm-up.
- Problem solving: Read each part of question 2 and give the children time to discuss it in their groups, before sharing with the rest of the class.
- Ask the children to draw a pictogram showing some data, such as the contents of the cutlery drawer at home. (They can make up their own quantities.) Ask them to make up questions about their pictograms similar to the ones in question 1. They should then exchange questions with a partner to answer each other's questions using their pictograms. For example:

Knives	
Forks	
Spoons	
Serving spoons	

- Continue with practical activities until you are sure the children can read and interpret pictograms, as well as collect and classify their own data and present it in a pictogram. There are many different topics, such as: favourite colours, fruits, songs, TV programmes, types of shoe worn, number of pieces of litter in different parts of the school, weather conditions on different days.

Support

If some children need extra practice with pictograms, use an empty pictogram frame and let them build up pictograms using small counters or objects of different colours. Give a mixed group of red, yellow, blue and green counters, and let them sort them into rows in a frame like this:

Colour	Number of counters
Red	
Yellow	
Blue	
Green	

> If there are any colour-blind children in your class, you could use other objects, such as coins of different values, or a mixture of shapes.

Answers for Pupil Book 2 page 57

1 **a** It uses pictures to show the number in each category.

    b 1 race won   **c** Kiara   **d** Navva

    e No, it only shows how many races they won.

Problem solving:

2 **a** favourite juice flavour   **b** 5

More pictograms

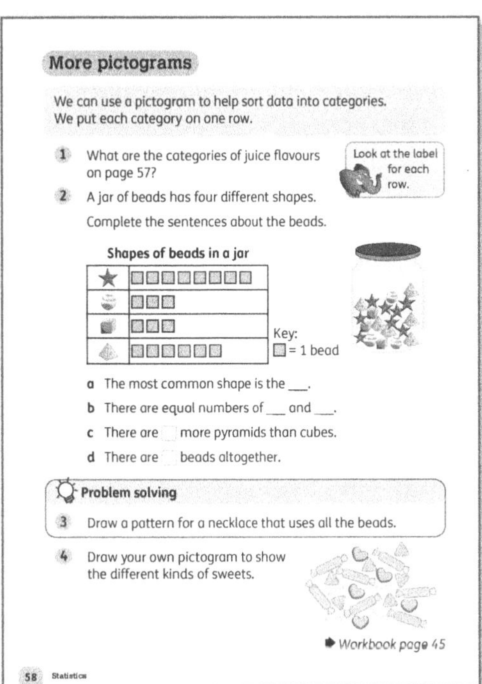

Warm-up

- Start the lesson by showing the class the pictogram of juice flavours from the previous lesson and asking them questions about it.
- For question 1 on **Pupil Book 2 page 58**, discuss the term *category*. Ask the children which categories of juice are shown in the pictogram. Some children may include the additional category (orange).

Focus

- You may choose to work through questions 2–4 with the class or read the questions and then let them answer independently.
- Spend some time talking about how many of each shape is shown in question 2. Let the children identify which rows have most beads before they count. Ask why it is easy to see this, in order to draw their attention to the way we read pictograms visually – the longer line of pictures easily shows us which has the most.
- Problem solving: Question 3 is very open ended, with scope for practical work, extension and mathematical thinking. You may choose to give the children objects or counters to work with, or to let them draw the shapes and work out their patterns by drawing.
- Discuss with the class how they would go about making their designs. What could they use to plan their design? Some children may suggest working practically, breaking the groups into equal amounts, then sharing out the leftover shapes. Others may choose to work numerically, using multiplication or division. Draw their attention to the way we could use symmetry to help build up the pattern. Some may work from one side to the other, and others may work from the middle outwards.
- For question 4, discuss with the class how we might present the data in the pictogram. Ask:
  - *How many categories do we have?'*
  - *How should we show this in a pictogram?*
  - *What symbol should we use for each sweet?*
  - *Can someone draw the key for me?*
- You might let the children suggest ways of drawing the pictogram on the board or, if they need assistance, draw up an empty frame and invite individual children to fill in the kind of sweet in the first column and the symbols in the second column:

Kind of sweet	Number of sweets

Follow-up

You may choose to work through **Workbook 2 page 45** with the class or let them work through it independently.

Answers for Pupil Book 2 page 58

1 mango, pineapple, apple, orange

2 a star   c 3
   b spheres and cubes   d 20

Problem solving:

3 Individual's answers

4 Possible answer:

Pink	● ● ● ● ●
Green	● ● ● ● ● ●
Yellow	● ● ● ● ● ● ●
Blue	● ● ● ●

Key: ● = 1 sweet

Answers for Workbook 2 page 45

1 a 4   b 2   c paints
   d chalks and wax crayons

2 Individual's answers

Block diagrams

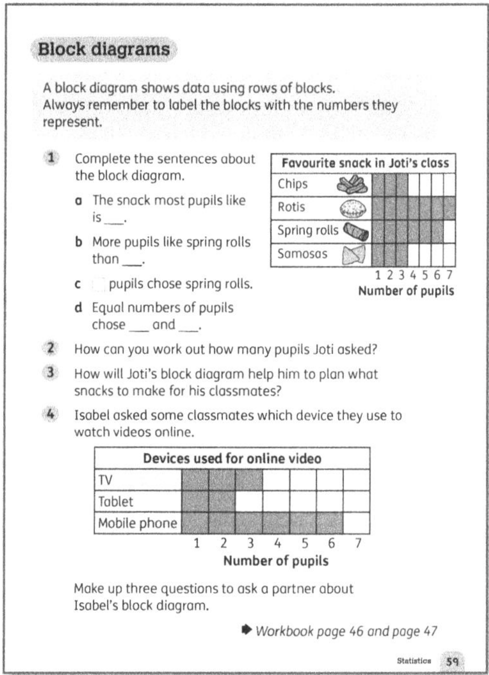

Materials

Block diagram and pictogram showing the Favourite day of the week from the 'Unit introduction' (page 87)

Warm-up

Show the children the 'Favourite days of the week' block pictogram and the block diagram you drew in the 'Unit introduction' (page 87).

Favourite day of the week

Monday
Tuesday
Wednesday
Thursday
Friday
Saturday
Sunday

Ask questions about the block diagram and pictogram:

* *What does the block diagram tell us?*
* *How is it different from the pictogram?*
* *What does this block diagram use instead of pictures?*
* *Is it easy to see which day the most children chose?*
* *Is it easy to see which days have equal numbers of votes?*
* *What makes it easy to see?*

Focus

Work through questions 1–4 on **Pupil Book 2 page 59** with the class. In question 4, the children can work in pairs to make up questions about the block diagram for their partner to answer.

Follow-up

Workbook 2 page 46 provides practice with collecting, recording and displaying data in a table and a block diagram. Let the children discuss in pairs how they would gather the information. Then let them ask ten different classmates to answer the question. They record the answers in the table. Afterwards, they put the data into a block diagram.

Answers for Pupil Book 2 page 59

1   **a** rotis   **b** chips or samosas   **c** 6
    d chips and samosas

2 Add together the number of blocks in the whole diagram: 19.

3 Possible answers: He knows to make more rotis and spring rolls than chips and samosas; He will make rotis because most pupils like them; He will make 3 portions of chips, 7 rotis, 6 spring rolls and 3 samosas so that everyone gets their favourite snack.

4 Possible answers:
* Which device did most pupils use?
* Which device did fewest pupils use?
* How many more pupils used mobile phone than tablet?
* How many children used TV/tablet/phone?

Answers for Workbook 2 page 46

1 and **2** Individual's answer

End-of-unit check

Use **Workbook 2 page 47** questions 1 and 2 to assess the children's understanding of representing data.

In addition, you can show the children a variety of graphs from this unit and ask questions such as:

* *Which of these graphs are pictograms* (show the different representations in the unit)? (Pictograms on **Pupil Book page 57** and **Workbook 2 page 45**.)
* *How do you know?* (Pictograms have a key and use a symbol to represent data, some children may recognise the similarity between pictograms and block diagrams.)
* *What is the difference between a pictogram and a block diagram?* (A block diagram uses filled or shaded blocks rather than pictures or symbols.)
* *Can you point to the top row in this representation* (show a pictogram or block diagram)? (For example, child points to the top row in the pictogram or block diagram.)
* *How many columns does it have?* (For example, child identifies the number of columns in the pictogram or block diagram being shown.)
* *What does the title tell us?* (The title explains what the diagram represents.)

Answers for Workbook 2 page 47

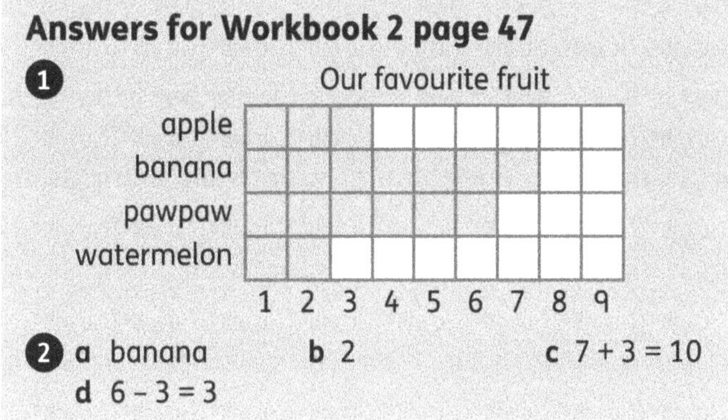

2   **a** banana   **b** 2   **c** 7 + 3 = 10
    d 6 − 3 = 3

Learning objectives

- Recall and use multiplication facts for the 2, 5 and 10 multiplication tables, including recognising odd and even numbers.

- Calculate mathematical statements for multiplication within the multiplication tables and write them using the multiplication (×) and equals (=) signs.

- Solve problems involving multiplication, using materials, arrays, repeated addition, mental methods and multiplication facts, including problems in contexts.

- Show that multiplication of two numbers can be done in any order (commutative).

Key words

multiply   many times   groups   equal   repeated addition   times   skip count   product   multiplication sign   array   times tables

Unit introduction

Materials

Counters; 1c and 5c coins (or similar); pictures showing groups of one to five objects

It is a very good idea to build up a picture collection of items that we regularly see in groups, sets, packs or pairs. Once you start looking out for these, you will see them in many different contexts. A useful lesson warm-up is to show some of these pictures and ask prompt questions such as: *How many ___ can you see?* or *How is the number arranged?*

You can use a smartphone to take photographs of items such as:
- eggs in boxes
- fruits or vegetables that are sold in twos, threes or fours
- bunches of bananas
- packs of two, three or four batteries
- flowers that have a clear 5-petal structure.

Most smartphones will allow you to build up albums of photographs that you can regularly use as a teaching prompt for a number talk or lesson opener.

Teaching guidance

- Revise counting in twos using a number line.
- Organise the children into groups and give each group several pictures showing 2 items (such as a pair of socks, 2 sweets, a pair of shoes, 2 fish) and some counters. Ask them to work out how many of each item there would need to be to give each member of their group 2. Let them find their own methods of working this out. Discuss as a class how they worked out how many items they would need.
- Repeat with pictures of 3 items. This time, make sure there are no more than 6 childrenin each group so that you do not go beyond 20 items.
- Give the children a number of drawing activities. For example, ask them to draw 4 groups of 2 birds, 5 groups of 3 socks, 8 groups of 2 sweets. Ask them to write the total number of objects in each set of groups.
- Use these activities involving money. (You can adapt to use coins from your currency or use play money.)
  ○ Practise counting 1c coins in twos.
  ○ Ask the children to complete a chart using 5c coins: one 5c = ☐ ; two 5c = ☐ ; three 5c = ☐ ; four 5c = ☐ . Do not go beyond 20c.
- Ask the children to make *groups* of 2 and 3 objects and to describe what they have made and done. When you are satisfied that they understand the concept of *repeated addition*, move on to groups of 4 and 5 objects. Reduce the number of children in each group to keep the number of items lower than 20.
- You can also use 'Target number calculations' (page 29) and 'Target number instructions' activities (page 30).

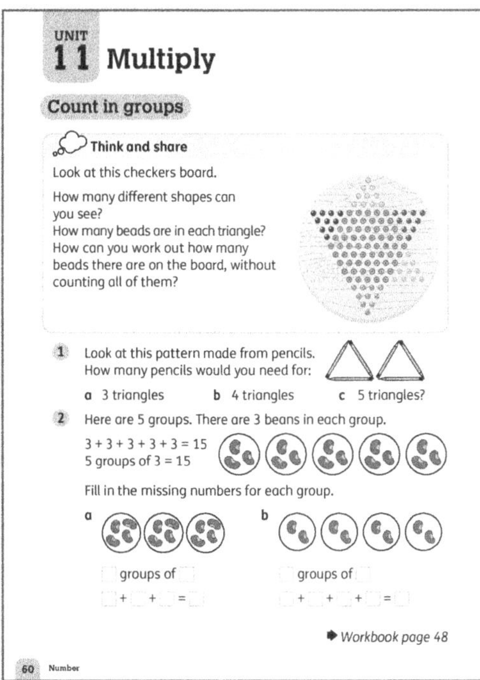

Warm-up

Think and share: Discuss the picture of the checkers board on **Pupil Book 2 page 60** with the children. Ask questions to prompt discussion such as:

- *What shapes do you see on the board?*
- *How would you draw these shapes?*
- *How are the beads arranged?*
- *How would you add up the beads in one triangle?*
- *What interesting patterns do you notice?*

Focus

Discuss the pattern made from pencils in question 1. Ask: *What is each shape made out of? How many pencils would we add to make a new triangle?*

The children can work in pairs to answer question 2 about the groups of beans.

Follow-up

The children can complete **Workbook 2 page 48**, independently as far as possible.

Answers for Pupil Book 2 page 60

Think and share: triangles, a hexagon, a large star shape, a large circle (the board) and small circles (the beads)

There are 10 beads in each triangle.

To work out how many beads there are on the board, you could repeat add tens. You could also multiply 10 by 6 (the number of triangles): $10 \times 6 = 60$.

1 a 9 b 12 c 15

2 a 3 groups of 4 $4 + 4 + 4 = 12$
  b 4 groups of 2 $2 + 2 + 2 + 2 = 8$

Answers for Workbook 2 page 48

1 a 6 groups of 2; 4 groups of 3; 3 groups of 4, 2 groups of 6

  b 10 groups of 2; 5 groups of 4; 4 groups of 5; 2 groups of 10

  c 15 groups of 2; 10 groups of 3; 6 groups of 5; 5 groups of 6; 3 groups of 10; 2 groups of 15

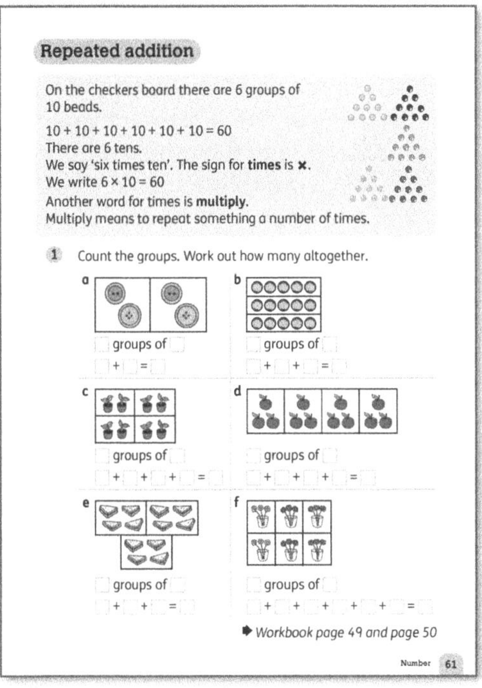

Materials

Counters or other small objects; real-life images of items in pairs or multipacks (optional)

Warm-up

Give the children counters or other small objects to repeat add. You can give them tasks such as:

- *Make 3 groups of 2. How many counters are there altogether?*
- *Now make another group of 2. How many groups have you got now? How many counters have you got?*
- *Let's add another group. There are now 5 groups of 2. How many counters are there altogether?*
- *Now make 2 groups of 5. What do you notice?*

Focus

- Work through the example on **Pupil Book 2 page 61** with the class. Point out that the word *multiply* means *times* and that the *multiplication sign* is another way of showing *times*.
- As you work through question 1, say that each of the pictures, as well as being described as 'groups of' and as a repeated addition, then can multiply to show how *many times*. Write each part (a–f) on the board as a multiplication sentence.

Follow-up

Use **Workbook 2 page 49 and page 50** to consolidate this work. Demonstrate an example from each page. The children can then complete the pages independently as far as possible.

Challenge

Ask: *How many different ways can you arrange counters in groups to make these totals?*

a 20   **b** 24   **c** 30

Support

- At this stage, the use of visuals to reinforce what is being calculated is very powerful. It also helps children to see whether they have solved problems correctly or not. Make sure you use language that relates to the problem at hand. For example, if you are talking about birds' legs, keep referring to *two legs*. Be careful of moving to abstract terminology too quickly, as this may confuse the children.
- Give more examples of groups of twos, threes, fours and fives. You may also wish to show real-life pictures of items in pairs or multipacks.

Answers for Pupil Book 2 page 61

1 **a** 2 groups of 2; $2 + 2 = 4$
   b 3 groups of 5; $5 + 5 + 5 = 15$
   c 4 groups of 2; $2 + 2 + 2 + 2 = 8$
   d 4 groups of 3; $3 + 3 + 3 + 3 = 12$
   e 3 groups of 4; $4 + 4 + 4 = 12$
   f 6 groups of 3; $3 + 3 + 3 + 3 + 3 + 3 = 18$

Answers for Workbook 2 page 49

1 **a** 2 groups of 3 = 6; $2 \times 3 = 6$
   b 3 groups of 4 = 12; $3 \times 4 = 12$
   c 3 groups of 3 = 9; $3 \times 3 = 9$
   d 6 groups of 2 = 12; $6 \times 2 = 12$
   e 2 groups of 5 = 10; $2 \times 5 = 10$
   f 3 groups of 5 = 15; $3 \times 5 = 15$

Answers for Workbook 2 page 50

1 **a** 2   **b** 4
   c 6, 8, 10, 12, 14, 16, 18, 20
2 **a** 2, 2, 2   **b** 4, 4, 4   **c** 6, 6, 6
   d 8, 8, 8   **e** 10, 10, 10

Arrays

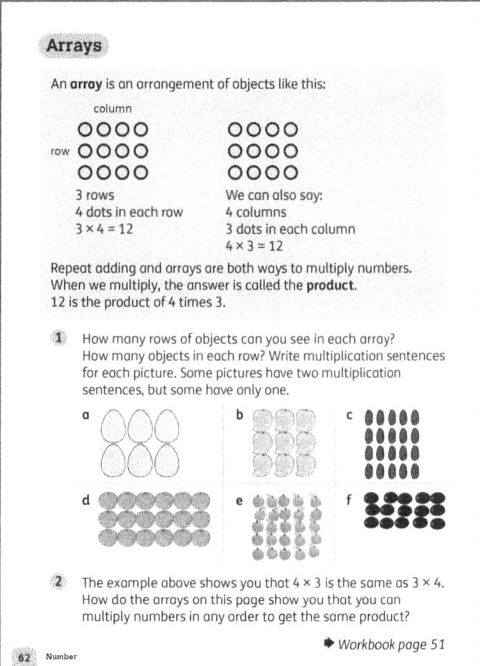

Materials

A variety of egg boxes, fruit trays or chocolate boxes – any box that suggests obvious arrays of 6, 8, 12 or 20 objects (you could also make grids of varying sizes); beans, counters and other objects to set out in arrays; number lines

Warm-up

Ask the children to arrange beans, counters or similar in sets of 2 and sets of 10. They then combine the sets with others and add up the number of beans in 1 set, 2 sets, and so on.

Focus

- Invite the children to build *arrays* with beads to prove that 2 sets of 4 beads is *equal* to 4 sets of 2 beads. They could then find other similar examples.
- Use arrays to begin to formulate facts for the 3 and 4 *times tables*:
  1 row of 3 = $1 \times 3 = 3$
  2 rows of 3 = $2 \times 3 = 6$, and so on.
- Remember to match the models you are using to the correct notation (calculation):
  2 groups of 3 is 2×3
  3 groups of 2 is 3×2
- Tell the children to make 10 groups of 2 counters. Ask them how many counters are in 1 group of 2, 2 groups of 2, and so on. Write the total on the board each time. Ask the children if they can see a pattern. Get them to make 10 groups of 10 counters. Ask the class how many counters are in 1 group of 10, 2 groups of 10, and so on. Write the total on the board each time. Ask them again whether they can see a pattern.
- Reinforce the pattern obtained by counting in tens using

a number line. Show the children that, starting at 0 and jumping 10 each time, they get the same pattern of numbers as when they added the sets of 10 together. Summarise the 10 times table with a whole-class session.

- Introduce the term *product* as what we get when we multiply as you work through the examples and questions 1 and 2 on **Pupil Book 2 page 62** with the children.

Follow-up

Work through **Workbook 2 page 51** with the children. They will do further work on the 2 times table throughout this unit.

Answers for Pupil Book 2 page 62

1 a 2 rows of 3 $2 \times 3 = 6$ $3 \times 2 = 6$
   b 3 rows of 3 $3 \times 3 = 9$
   c 4 rows of 5 $4 \times 5 = 20$ $5 \times 4 = 20$
   d 3 rows of 6 $3 \times 6 = 18$ $6 \times 3 = 18$
   e 5 rows of 5 $5 \times 5 = 25$
   f 3 rows of 5 $3 \times 5 = 15$ $5 \times 3 = 15$

2 It doesn't matter whether we add up the columns or the rows in an array; the total is the same.

Answers for Workbook 2 page 51

1 a $4 \times 2 = 8$; $2 \times 4 = 8$ [Provided as an example]
   b $4 \times 3 = 12$; $3 \times 4 = 12$ c $1 \times 4 = 4$; $4 \times 1 = 4$
   d $2 \times 5 = 10$; $5 \times 2 = 10$ e $5 \times 1 = 5$; $1 \times 5 = 5$
   f $5 \times 4 = 20$; $4 \times 5 = 20$

Multiply by 0 or 1

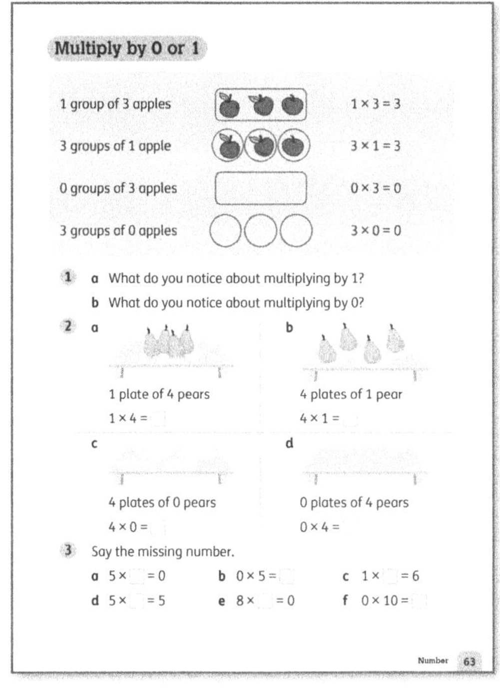

Materials

Paper plates, dishes or cups for making groups of objects; counters, large beads or small objects for making equal sets

Warm-up

- Set out 2 plates, with 3 objects on each plate. Explain to the class that you are going to make sets of 3 objects. Ask:
  - *How many sets of 3 do I have?*
  - *How many objects do I have altogether?*
  - *How can we write this as a multiplication sentence?*
- Clear the plates and objects and explain that you are going to use a different number now. Set out 1 plate with 4 objects. Ask:
  - *How many plates are there?*
  - *How many items are on each plate?*
  - *How can we write this as a multiplication sentence?*

Focus

- Use the plates of objects to demonstrate how we make, for example, '1 group of 4'. You can also then set out the individual objects to demonstrate '4 groups of 1'. Use the terms *group* and *equal* to emphasise the idea that a single group of a given number is always equal to the number. Once you can see that the children understand this, make sure you translate it to show the multiplication sentence.
  $4 \times 1 = 4$
  $1 \times 4 = 4$
- Then ask: *How many we would have if we make 0 groups of 4?* Talk about this in real-life terms, for example: *Imagine that a pack has 4 batteries. I look on the shelf and there are no packs – zero. 0 groups of 4 equals how many? A plate has 4 cookies. I have no plates left. How many cookies do I have?*
- Also represent this as multiple groups of zero: *If I have 4 plates of 0 cookies, how many cookies do I have?*
- When you can see that children understand that 0 groups of a given number (or any number times 0) is always equal to 0, show the multiplication sentence. For example:
  $4 \times 0 = 0$
  $0 \times 4 = 0$
- Read through the examples on **Pupil Book 2 page 63** before working through questions 1–3 with the children. Some children may be able to work independently.

Challenge

Give problems involving addition or subtraction of groups, for example:

- *I have 5 plates with 3 cookies on each plate. I take away the 5 plates. How many cookies are left?*
- *A bunch of bananas has 3 bananas. I have 0 bunches. I get 3 more bunches. How many bananas do I have?*
- *I have 1 pack of 10 crayons. I give it away. How may packs do I have? How many crayons do I have?*

Interesting mistakes

- '$5 \times 1 = 1$' is a mistake you may see when children apply the 'rule' for multiplying by 1 incorrectly. Remind them always to think in terms of real objects – 5 sets of 1 object, or 1 set of 5 objects.

- Similarly, '0 × 1 = 1' is a mistake that comes up when children mix up the rule for multiplying by 1 with the rule for multiplying by 0.

Answers for Pupil Book 2 page 63

1 **a** Multiplying a number by 1 gives you the same number.
   b Multiplying a number by 0 gives you 0.

2 **a** 4   **b** 4   **c** 0   **d** 0

3 **a** 0   **b** 0   **c** 6
   d 1   **e** 0   **f** 0

The 2 times table

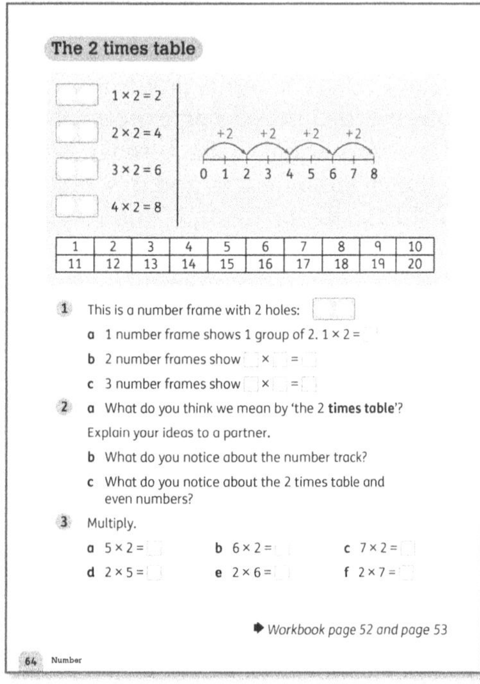

Materials
Number lines; 2c coins (or similar coins in a different currency, or play coins with 2 on them); objects for counting

Warm-up
Remind the children of the term *skip count*. They have already learnt how to skip count in twos and will have also worked on the 2 times table. Revise this work by skip counting in twos on a number track and by inviting the children to arrange objects in pairs and count them in twos.

Focus
- Ask the children whether they remember the terms *odd* and *even*. Ask them to identify odd and even numbers on the number track. Ask: *What do you notice about the numbers in the 2 times table?* (Numbers are always even when we multiply by 2.)
- Give the children sets of 2c coins to add up. If there are coins with a value of 2 in your currency, use those instead or use play money.

- Work through the information at the top of **Pupil Book 2 page 64** together before completing questions 1–4 with the children.

Follow-up
- Use **Workbook 2 page 52** to consolidate work on the 2 times table. Most children should be able to work through this independently once you have explained the instructions.
- Revise the words odd and even before the children work through **Workbook 2 page 53**.

Challenge
Children who finish quickly could be given this problem to solve:

A school bus has many rows, each with 2 seats. But the front row and the back row each have 1 seat.
- *Does the bus have an even or odd number of seats? Explain how you know.*
(The bus has an even number of seats, because we can add together the front and back seat to make a pair, and the rest are all pairs too.)
- *If there are 8 rows with 2 seats, how many seats are there altogether in the bus?*
(There are 18 seats in total.)

The children can also investigate rules about odd and even numbers.

Give each of the statements below one at a time and ask:
- *Is this statement true or false? If it is true, why is it true? Can you find a way to draw or show that it will always be true?*
- *If it is false, why is it false? Can you find a way to draw or show that it will always be false?*

Statements:
- *If you add an even number to an even number, the answer is always even.*
- *If you add an odd number to an odd number, the answer is also always even.*
- *If you multiply an even number by an odd number, the answer will also be even.*
- *If you multiply an odd number by an odd number, the answer will be odd.*

Answers for Pupil Book 2 page 64

1 **a** 2   **b** 2 × 2 = 4   **c** 3 × 2 = 6

2 **a** All the numbers multiplied by 2
   b All the numbers in the 2 times table/all the even numbers are shaded.
   c They are the same numbers.

3 **a** 10   **b** 12   **c** 14
   d 10   **e** 12   **f** 14

Answers for Workbook 2 page 52

1   a   2, 2   b   4, 4   c   6, 6   d   8, 8
    e   10, 10   f   12, 12   g   14, 14   h   16, 16
    i   18, 18   j   20, 20

2   Possible answer: Multiply the tens and the ones by 2 separately (10 × 2 to make 20 and 1 × 2 to make 2) and then add these two numbers to get the answer, 22.

Answers for Workbook 2 page 53

1   a   10, even   b   11, odd   c   13, odd
    d   20, even   e   17, odd   f   21, odd

2   a   even   b   even   c   odd   d   odd
    e   odd   f   even   g   odd   h   odd

3   99 is odd. I know because the last digit is odd.

Multiply by 5 or 10

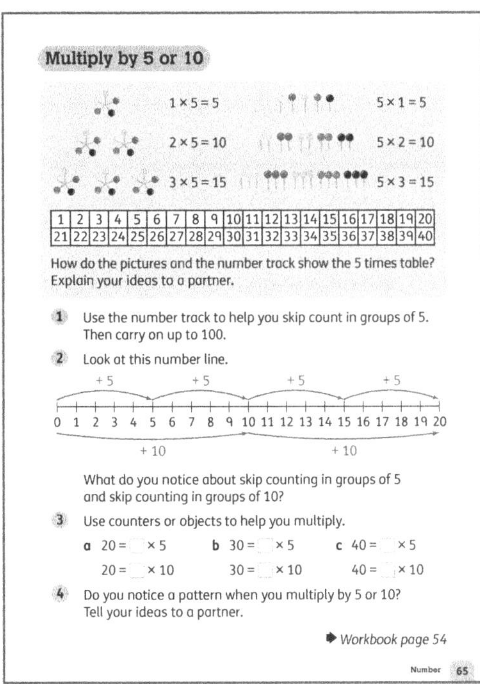

Materials

Objects for counting and modelling (such as plates, trays or dishes for making sets of objects; counters or objects for counting; or, if possible, toothpicks and marshmallows or modelling clay); large sheets of paper for making posters (optional)

> The models at the top of **Pupil Book 2 page 65** demonstrate a highly effective way of modelling the *commutative* property of multiplication: the idea that 1 × 5 is 1 group of 5, and that 5 × 1 is 5 groups of 1, and that both calculations have the same answer.

Warm-up

- The children can talk in pairs about what they can see in the pictures at the top of **Pupil Book 2 page 65** and on the number track. Each diagram has 5 components: the models on the left-hand side show sets of 5 and the models on the right break up each set of 5 into ones.
- Give the children objects to model these sets. Ask the children to make the sets in the diagrams and setout the different multiplication sentences using fives or ones.

Focus

- Work through questions 1–4 using the children's models as well as the number track and number line on **Pupil Book 2 page 65**.
- In question 3, encourage the children to recognise that all numbers in the 10 times table are even whereas in the 5 times table the numbers alternate between odd and even numbers.

Follow-up

The children complete **Workbook 2 page 54** independently, as far as possible.

Challenge

Ask: *What relationship do you notice between multiplying by 10 and multiplying by 5?*

Support

- Use the models to help the children find ways of modelling the 2 times and 3 times tables. If needed, use the models to help you build up posters of the different times tables for your classroom.
- Use geoboards to help the children model the different times tables using arrays: 1 row of 5, 2 rows of 5, 3 rows of 5, and so on.

Answers for Pupil Book 2 page 65

1   5, 10, 15, 20, 25, 30, 35, 40, 45, 50, 55, 60, 65, 70, 75, 80, 85, 90, 95, 100

2   Possible answers: When we skip count in twos, each jump is double the jump made when we skip count in fives.

    When we skip count in tens we get the same numbers as when we skip count in fives, but with every other number missing.

3   a   4, 2   b   6, 3   c   8, 4

4   Multiplying by 5 always gives numbers that end in 5 or 0. Multiplying by 10 always gives numbers that end in 0.

    Other possible answers: We can halve the multiples of 10 to get the multiples of 5. We can double the multiples of 5 to get the multiples of 10.

    The children may also mention patterns in the tens digits.

Answers for Workbook 2 page 54

1   a   15   b   30   c   25
     d   35   e   4   f   9

2   Possible answers:

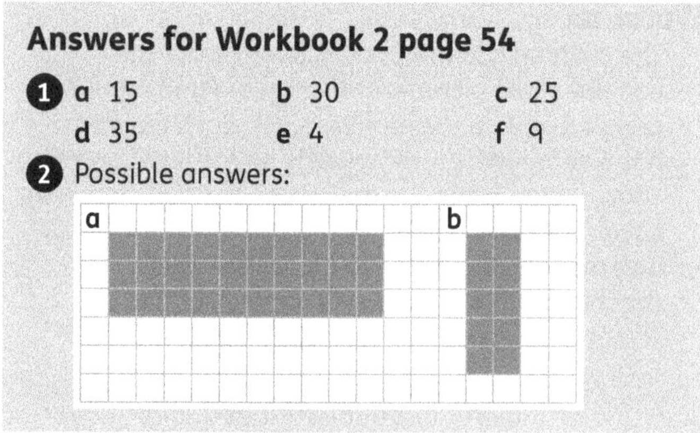

Answers for Pupil Book 2 page 66

1   a   12   b   12   c   6
     d   4   e   0

2   a   1   b   2

     c   Possible answers: The products in the 4 times table are half of the products in the 8 times table. The products in the 8 times table are double the products in the 4 times table.

     d   20, 40, 60, 80

     e   numbers on the number track from the 3 times table: 12, 24, 36, 48, 60, 72

       Possible answer: Add the digits of each product in the 3 times table show a pattern of 3, 6 and 9.

Problem solving:

3   5 packets

Multiplication practice

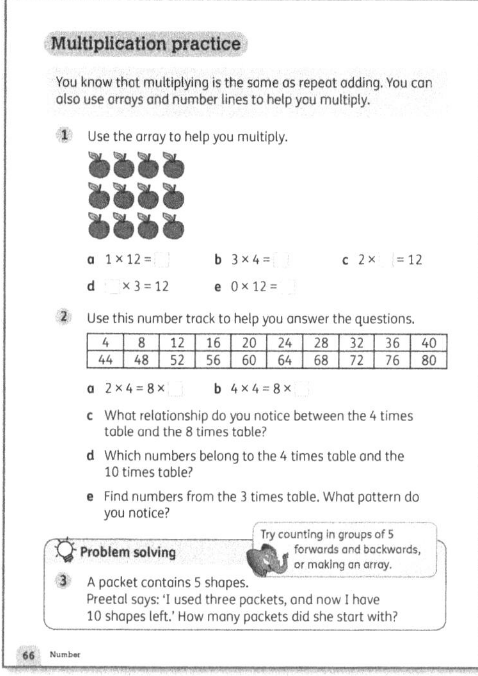

Solve multiplication problems

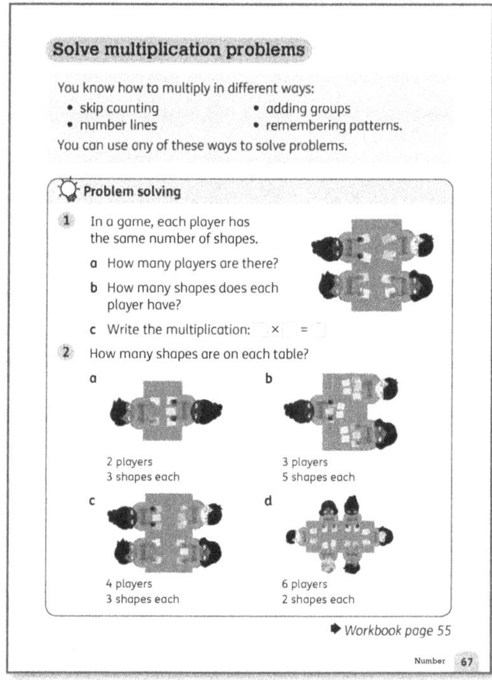

Materials

Number lines; objects for creating arrays

Warm-up

Practise the 2, 4 and 8 times tables as a class. Build each multiplication as an array on the board to get the children thinking about and visualising patterns in the times table.

Focus

- Present question 1 and question 2 on **Pupil Book 2 page 66** as number talks. (see 'Number talks', pages 17–18).
- Problem solving: Then let the children work independently in pairs on question 3. Assist children who need extra support.

Materials

Number lines; objects for creating arrays

Warm-up

Quickly revise each of the methods for multiplication listed at the top of **Pupil Book 2 page 67**: skip counting, number lines, adding groups, remembering patterns.

Problem solving can cause anxiety in children if they feel they have to find the answer quickly. When we take the pressure off children to work fast, and we shift the emphasis from getting the correct answer to the process of thinking about 'what's going on', we allow space for them to work slowly and methodically towards developing their mathematical reasoning.

Focus

<u>Problem solving:</u> The children can work in pairs or small groups to answer question 1 and question 2 on **Pupil Book page 67**. Remind children to consider which strategy they can use to solve the problems.

Follow-up

Go through the questions on **Workbook 2 page 55** and discuss them with the class before allowing the children to complete the problems as independently as possible.

Support

Some children find it difficult to work out how to translate a word problem into a multiplication sentence. Spend some time asking them questions so that they can work out what the problem requires them to do, and how to do it. You may find it useful to start with a numberless problem ('Numberless word problems' page 18), such as:

Some children are going to play a game. Everyone is going to get the same number of shapes.
- *What questions can we ask about the game?*
- *How do we know how many shapes we need?*
- *What do we need to find out?*

Answers for Pupil Book 2 page 67

<u>Problem solving:</u>

1 a 4   b 2   c 4 × 2 = 8
2 a 6   b 15
   c 12   d 12

Answers for Workbook Book 2 page 55

1 a 7 groups of 2 = 14   b 3 groups of 3 = 9
   c 9 groups of 4 = 36   d 3 groups of 2 = 6
   e 5 groups of 5 = 25

End-of-unit check

Ask questions and carry out practical activities to assess the children's understanding of different types of multiplication, for example:
- Set out an array (for example, with 3 rows of 4 counters) and ask: *How many rows are there?* (3) *How many columns?* (4) *How can we work out how many there are altogether?* (Multiply 3 by 4 or 4 by 3 to give 12.) Repeat with other arrays.
- Ask the children to count in twos, threes or fives. (2, 4, 6, 8, 10, 12, 14, 16, 18, 20; 3, 6, 9, 12, 15, 18, 21, 24, 27, 30; 5, 10, 15, 20, 25, 30, 35, 40, 45, 50)
- Ask: *Starting at 20, if you make 3 jumps of 10 along a number line, what numbers do you land on?* (20, 30, 40, 50.)
- Ask questions such as: *What is 2 × 3?* (6) *What is 6 × 3?* (18) *What is 4 × 10?* (40)
- Ask: *How many groups of 10 are needed to make 40?* (4)
- Challenge the children to write another multiplication sentence that has the same answer as 2 × 5 or 4 × 3, for example. (For example, 5 × 2; 1 × 10; 10 × 1, 3 × 4; 2 × 6; 6 × 2; 1 × 12; 12 × 1)

Use the 'true or false' strategy to test the children's understanding of vocabulary and their ability to apply calculation skills. Give the children a statement and ask them to say whether it is true or false. Discuss how they decided. Here are some examples:

True or false?
- *1 multiplied by 7 is 8.* (false)
- *2 times 3 is 6.* (true)
- *The difference between 6 and 15 is 9.* (false)
- *The sum of 3, 4, 6 and 2 is 15.* (true)
- *5 taken away from 40 is 45.* (false)
- *4 lots of 2 are 8.* (true)
- *The product of 5 and 7 is 35.* (true)
- *Half of 50 is 100.* (false)
- *9 times 10 is 19.* (false)
- *15 divided by 5 is 3.* (true)

Use the properties of shapes to reinforce multiplication and division facts, for example:
- Display a triangle and ask: *How many sides are in 2 (or 5 or 10) triangles?* (6; 15; 30)
- Say: *I have a number of triangles. There are 27 sides. How many triangles are there?* (9)
- Repeat with other shapes and numbers.

Learning objectives

- Recall and use multiplication and division facts for the 2, 5 and 10 multiplication tables, including recognising odd and even numbers.

- Calculate mathematical statements for multiplication and division within the multiplication tables and write them using the multiplication (×), division (÷) and equals (=) signs.

- Solve problems involving multiplication and division, using materials, arrays, repeated addition, mental methods, and multiplication and division facts, including problems in contexts.

- Show that multiplication of two numbers can be done in any order (commutative) and division of one number by another cannot.

- Understand division as: sharing (number of items per group); grouping (number of groups); repeated subtraction.

Key words

divide   equal groups   shared equally   fact family

Unit introduction

Teaching guidance

- Use a number talk to open this unit. By now you should be familiar with the procedure for number talks. (If not, see 'Number talks', pages 17–18.) Draw a group of 24 objects on the board. To introduce the meaning of *equal groups* and *shared equally*, ask: *What different ways can I use to share these objects into equal groups? I want all the groups to have the same number of objects.*
- Allow plenty of time for this number talk. The children should work out and explain their ways of dividing the objects into equal groups of different sizes.
- In addition, see 'Use shapes to reinforce multiplication and division facts' (page 30).

Repeated subtraction

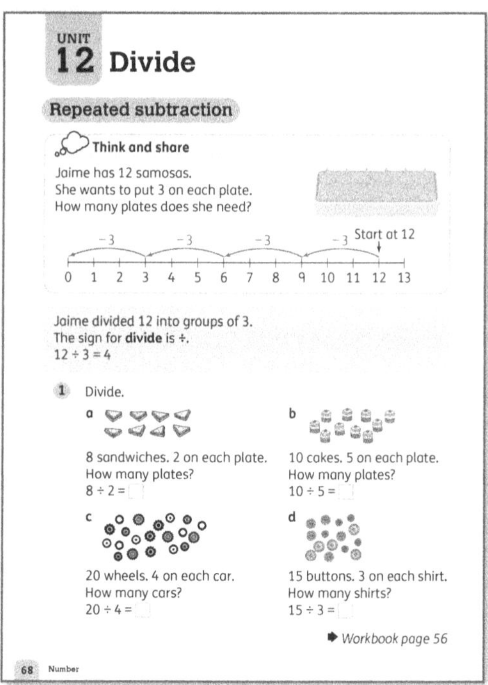

Materials

Objects for counting (a minimum of 28); counters and boxes (optional)

Warm-up

- Use the following numberless word problem as the basis of a number talk (see 'Numberless word problems' and 'Number talks' on pages 17–18): *I make a lot of cupcakes. I want to sell them in little boxes with the same number in each box. How can I work out how many boxes I will need?*
- Discuss the problem with the class, following the steps for a number talk. Then gradually feed in the following values: replace 'a lot of cupcakes' with '28 cupcakes'. Replace 'the same number' with the number 4.
- Go through the steps again, letting the children suggest ways to work the problem out. Note that they may employ a number of different methods, depending on their understanding of multiplication, addition and subtraction. Focus on getting them to share their thinking and reasoning and on deciding whether their reasoning works.

Focus

- If repeated subtraction comes up in the discussion, you can illustrate the points on the board as it arises. If not, you may want to focus on the following way of working.

- Make a group of 28 objects. Count them with the class. Then say: *Let's say each object is one cupcake. Now I want to see how many times I can take four away until I have used them all up.* Invite the children to guess how many groups of 4 you will be able to take away. Note their guesses in a corner of the board.
- Take away groups, noting the repeat subtractions:
  $28 - 4 = 24$ $24 - 4 = 20$ $20 - 4 = 16$
  (1 group) (2 groups) (3 groups)
- Continue in this way until you have shown all 7 groups subtracted. Explain that this is a way of working out how many equal groups we can take away.
- <u>Think and share:</u> Look at **Pupil Book 2 page 68** with the class. Discuss what is different and what is the same about this way of showing the repeat subtraction. (The Pupil Book shows a number line with jumps back, each jump showing a subtraction.)
- Work through several more examples in this way before you look at the problems in question 1 with the class.

Follow-up
- **Workbook 2 page 56** presents an alternative way of working. Instead of repeatedly subtracting a given equal number, we subtract one per box, building up the equal groups.
- Work together through the example at the top of the page, which shows six counters and three boxes. On the board, show how we might do this with repeat subtraction, subtracting 3 at a time, to get the same answer.
- Let the children work independently through the rest of the page. If they wish, they can use counters and boxes to model the questions in the Workbook.

Challenge
Ask the children to explain what multiplication sentence could be used to add up the trays in the Workbook and make the total number of counters again.

Answers for Pupil Book 2 page 68

<u>Think and share:</u> She needs 4 plates.

❶ a 4   b 2   c 5   d 5

Answers for Workbook 2 page 56

❶ a 2 in each box   b 5 in each box
  c 4 in each box   d 3 in each box

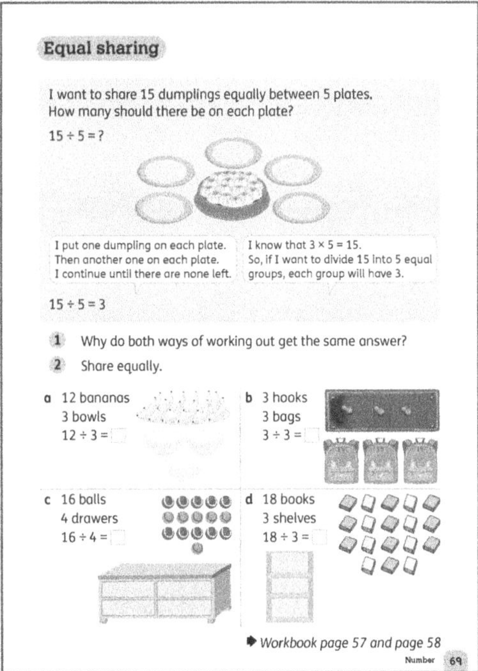

Materials
Objects or counters for sharing into groups

As in the unit introduction, the children can continue to explore different ways of dividing a number into equal-sized sets. The numbers 12, 16 and 36 will provide the children with lots of different ways to divide. Work with concrete objects if you need to, but you can also give the children marked number lines for them to investigate grouping in different ways.

Warm-up
- Present the example from **Pupil Book 2 page 69** as a number talk and work through it together (see 'Number talks', pages 17–18)

Focus
- Discuss question 1 with the class.
- The children can work independently through question 2 or you may want to work through these problems with the class.

Follow-up
- On **Workbook 2 page 57**, the children colour the shares of chocolate. Go through the first part of question 1 with them and then let them work independently to complete the remaining parts.
- Similarly, for **Workbook 2 page 58**, make sure the children understand the questions before allowing them to complete the page.
- Introduce the idea of dividing an amount into equal groups and use the division sign to represent this. Rewrite the different ways of sharing 12 equally using the division sign: $12 ÷ 2 = 6$, $12 ÷ 3 = 4$, $12 ÷ 4 = 3$ and $12 ÷ 6 = 2$.

- The children may initially find it difficult to understand the meaning of the division sign. It will help if you use word problems that include the phrases 'make equal groups' or 'grouped into' before using the ÷ sign and the term *divide*. The children should become more familiar with this format by completing the Workbook question using the division sign.

Answers for Pupil Book 2 page 69

1 Both methods involve sharing 15 to make 3 equal groups.

2 a 4   b 1   c 4   d 6

Answers for Workbook 2 page 57

1 a 2 groups of 12 blocks
  b 3 groups of 8 blocks
  c 4 groups of 6 blocks
  d 6 groups of 4 blocks
  e 8 groups of 3 blocks
  f 12 groups of 2 blocks

Answers for Workbook 2 page 58

1 a $12 \div 2 = 6$ $12 \div 6 = 2$
    $12 \div 3 = 4$ $12 \div 4 = 3$
  b $20 \div 2 = 10$ $20 \div 10 = 2$
    $20 \div 4 = 5$ $20 \div 5 = 4$
  c $24 \div 2 = 12$ $24 \div 4 = 6$
    $24 \div 6 = 4$ $24 \div 8 = 3$

2 a 7   b 1

Use arrays

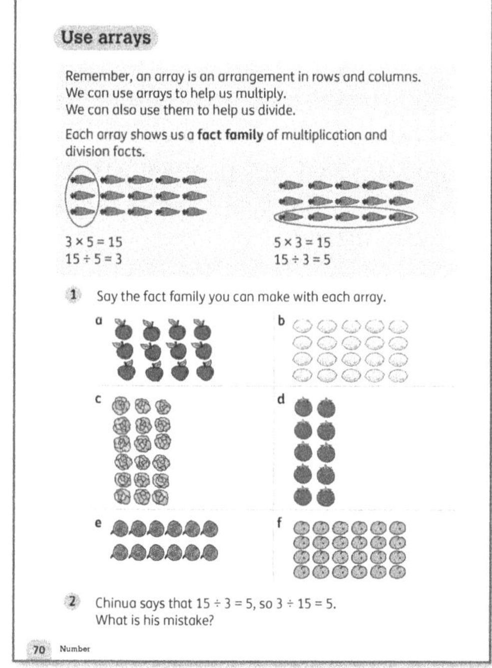

Materials

Photocopied or laminated grids for making arrays; small objects or counters; sheets of paper or cardboard for creating array and fact family charts

Warm-up

- The children have encountered arrays in multiplication already. They may also have already made connections between multiplication and division. Set out an array of 12 objects and ask them if they remember arrays from their previous learning. Remind them that they used arrays to multiply and remind them of the term *fact family*. Now ask: *How can I use this array to work out how many groups of 3 there are in 12?* (4) *How many groups of 4 are there in 12?* (3) *How many groups of 6 are there in 12?* (2)
- Demonstrate that we can build up rows of 3 to make 12 (4 rows of 3 make 12, or $4 \times 3 = 12$). That means that we can also break down the array, taking away rows of 3 to get back to 0.

Focus

- Emphasise that when we multiply a number, we build up groups; when we divide, we break down into groups.
- As a reminder, you could hand out sheets of paper and give each child a number. Ask them to create an array of rows and columns showing a fact family for that number. You can use these arrays and fact families as charts for your classroom.
- Work through the example and question 1 on **Pupil Book 2 page 70** with the class.
- For question 2, discuss with the children that, although we can multiply in any order, we cannot do the same with division. You can demonstrate this using shapes. In the example given, dividing 3 shapes into 15 equal parts would not result in 5 shapes in each group.
- The children may notice a similarity to addition and subtraction as addition can be done in any order, but subtraction cannot.

> Some children may realise that dividing 3 shapes into 15 equal parts would result in $\frac{3}{15}$, but we have not dealt with fractions yet, so that answer does not necessarily have to come up or be explained in detail.

Answers for Pupil Book 2 page 70

1 a $3 \times 4 = 12$ $4 \times 3 = 12$ $12 \div 4 = 3$ $12 \div 3 = 4$
  b $4 \times 5 = 20$ $5 \times 4 = 20$ $20 \div 4 = 5$ $20 \div 5 = 4$
  c $6 \times 3 = 18$ $3 \times 6 = 18$ $18 \div 6 = 3$ $18 \div 3 = 6$
  d $5 \times 2 = 10$ $2 \times 5 = 12$ $10 \div 5 = 2$ $10 \div 2 = 5$
  e $2 \times 6 = 12$ $6 \times 2 = 12$ $12 \div 2 = 6$ $12 \div 6 = 2$
  f $4 \times 6 = 24$ $6 \times 4 = 24$ $24 \div 4 = 6$ $24 \div 6 = 4$

2 You can multiply two numbers in any order to get the same total, but you cannot divide the same two numbers in any order: $15 \div 3$ is not the same as $3 \div 15$.
  Chinua should have said that $15 \div 3 = 5$, so $15 \div 5 = 3$.

Divide and multiply

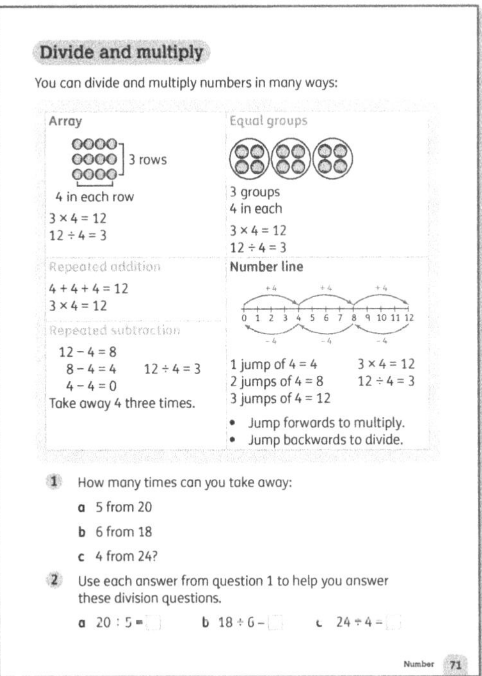

Materials
Counters or concrete objects; number lines

Warm-up
- This lesson revisits the various strategies that the children have used for multiplication and division. You can review any of these as the warm-up for this lesson.

Focus
- Discuss the different ways to multiply and divide numbers on **Pupil Book 2 page 71** as a class.
- Let the children work independently on question 1 and question 2. Encourage them to use one strategy to solve the problem and a second strategy to check their answer.

Challenge
Use the same approach discussed in Interesting mistakes but develop it further. You could use the examples given in the table below and encourage children to think about how they could replace each part of the sentence to make different story calculations.

Whole set	Number of groups	Number in each group
children in our class	number of groups of children	children in each group
a pen collection	number of pencil cases	pens per pencil case
a collection of counters	rows	columns
a bag of sweets	plates	sweets on each plate

Support
Some children need additional support to make the connection between multiplication and division, and between division and repeated subtraction (making groups). Lots of practice with concrete objects and number lines at this early stage will help them to consolidate the ideas and make it easier for them to understand the inverse nature of multiplication and division as they progress through the stages.

Interesting mistakes
3 ÷ 4 = 12, for example, is a mistake that results from confusion about the commutative law in multiplication, and the relationship between multiplication and division. Some children may be unsure about the order of the numbers in division. This explanation might help:

- *This is how we divide:*
  whole set ÷ smaller group size
  = number of groups we can make
  or
  whole set ÷ number of groups we want to make
  = number in each group.
  We always start with the whole set.
- *This is different from multiplication:*
  smaller group size × number of groups = whole set
  or
  number of groups × group size = whole set

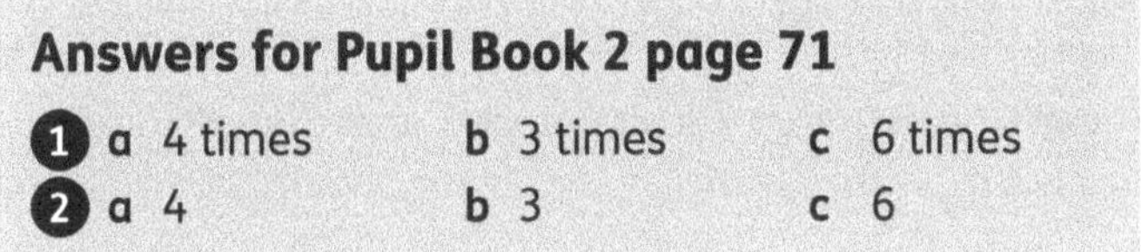

Answers for Pupil Book 2 page 71

1 a 4 times   b 3 times   c 6 times
2 a 4   b 3   c 6

Solve division problems

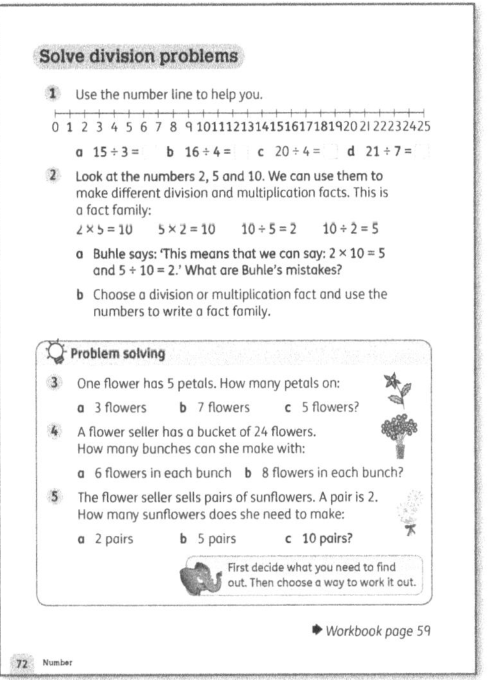

Materials
Counters (optional)

Warm-up

Prepare some function machines ('Function machines', page 30) with a focus on dividing for the children to complete. Provide counters for the children to model the divisions if you wish.

Focus

- Look at questions 1 and 2 on **Pupil Book page 72** as a class. Give the children time to discuss question 2a in pairs before discussing as a class.
- Problem solving: The children can work through questions 3–5 in pairs. Allow the children to select the strategy that makes the most sense to them. For each problem, they should first work out what is happening in the story and then what information they need to find. They should then choose a way to work it out.

Follow-up

Use **Workbook 2 page 59** to consolidate work on division and multiplication.

Answers for Pupil Book 2 page 72

1   **a** 5   **b** 4   **c** 5   **d** 3

2   **a** Possible answers: She has put the number in the wrong order in the multiplication.

In the division, she has assumed that $5 \div 10$ is the same as $10 \div 5$.

Problem solving:

3   **a** 15   **b** 35   **c** 25

4   **a** 4   **b** 3

5   **a** 4   **b** 10   **c** 20

Answers for Workbook 2 page 59

1   **a** $6 \times 3 = 18$ $18 \div 3 = 6$ $18 \div 6 = 3$

    b $5 \times 4 = 20$ $20 \div 4 = 5$ $20 \div 5 = 4$

    c $8 \times 3 = 24$ $24 \div 3 = 8$ $24 \div 8 = 3$

    d $10 \times 4 = 40$ $40 \div 4 = 10$ $40 \div 10 = 4$

End-of-unit check

Ask questions to assess the children's understanding of division, for example:

- *6 mangoes are shared equally between 3 people. How many mangoes does each person get?* (2)
- *How many sets of 5 can be made from 10?* (2)
- *There are 12 sweets in a bag. How many children would get 2 sweets each if they were shared equally?* (6)
- *12 sweets are shared equally between 3 people. How many sweets does each person get?* (4) *What if another person came along and wanted an equal share?* (3)
- *How many sets of 6 can be made from 24?* (4)
- *There are 16 samosas in a box. How many children would get 4 samosas each if they were shared equally?* (4)
- *How many groups of ... can I make from ...?*
- *What is ... divided by ...?*

UNIT 13 Fractions

Learning objectives

- Recognise, find, name and write fractions $\frac{1}{3}$, $\frac{1}{4}$, $\frac{2}{4}$ and $\frac{3}{4}$ of a length, shape, set of objects or quantity.

- Understand that fractions can act as operators (be interpreted as division).

- Write simple fractions for example, $\frac{1}{2}$ of 6 = 3 and recognise the equivalence of $\frac{2}{4}$ and $\frac{1}{2}$.

Key words

fraction   whole   part   equal   unequal

equivalent   half   halves   quarter

three-quarters   thirds

Unit introduction

Materials

Objects and pictures that can be cut into pieces; paper plates; flashcards showing half, quarter, $\frac{1}{2}$ and $\frac{1}{4}$; a sheet of paper for each child divided into half and a square divided into 4 quarters

Teaching guidance

- Teach the concept of a *whole* and *parts* to the class. The children have encountered this before, but it is necessary to revisit it.
- You can do this by showing a wide range of pictures and objects cut into pieces. Show the children that the pieces can be put together to make a whole. For example, show a picture of a house that you have previously cut into 3 pieces, or a slice of bread cut into 4 pieces.

- Explain what each picture shows. For example, say: *This picture of a house has been cut into 3 parts. We can put the parts together to make a whole. These 4 pieces of bread can be put together to make a whole slice.*
- Use the terms *equal* and *unequal* as you explain that sometimes the parts can be equal. Stress that this means they are the same size and that they will fit exactly on top of each other. Point out that unequal is the term for parts that are not equal. Demonstrate this using a sheet of paper. Fold it in half and show the children that the 2 halves are exactly the same size. Use the term *halves* as you do this can also cut a paper plate into 4 equal pieces to show them *quarters*.
- Give each child a sheet with a square divided into half, and a square divided into 4 quarters. Explain that the first square is divided into 2 equal parts. Revise the word *half* and teach the *fraction* notation $\frac{1}{2}$. Use flashcards to label half of a shape and display this in the classroom. Repeat this for a quarter.
- Get the children to colour half the square and label it. Repeat this for a quarter.
- Put the children into pairs. Ask each child to draw half a shape, then give their drawing to a partner. Their partner should draw the other half. Remind them that the 2 halves must be equal.
- Ask the children to draw a quarter of a cake. Their partner should draw another quarter of the cake. Again, stress that the quarters must be equal. Ask the children to label the parts of their completed drawings.

Whole and parts

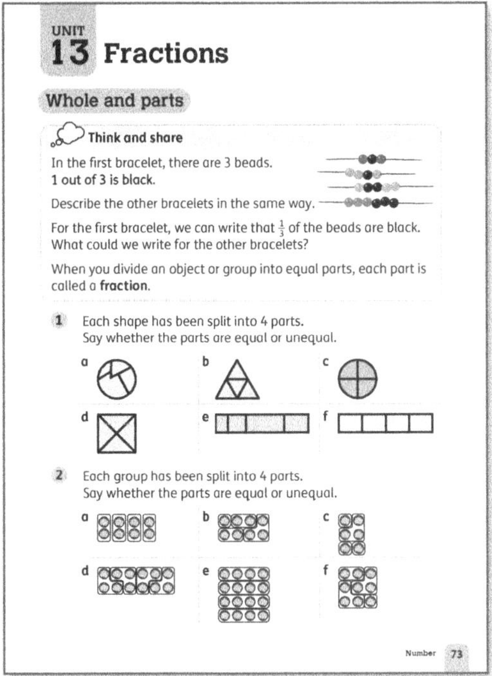

Materials

Paper 2D shapes (squares, rectangles, circles and triangles) that can be folded into halves and quarters

Warm-up

Give the children a set of 2D shapes. Let them investigate how to divide the shapes into halves and quarters by folding.

Focus

- <u>Think and share:</u> Discuss the picture of the bead bracelets on **Pupil Book 2 page 73**. Work through question 1 and question 2, discussing with the children:
  - the number of beads on each string
  - how many of the total number of beads are black
  - how to write the fraction of beads that are black for each one.
- Discuss which shapes in question 1 are divided into equal parts and which are divided into unequal parts.
- Discuss which groups in question 2 are divided into equal sets of circles and which are divided into unequal sets.

Challenge

Use the picture of the bead bracelets on **Pupil Book 2 page 73** as a starting point. Ask the children to make their own bead patterns to show fractions. Can they show $\frac{1}{3}$, $\frac{2}{3}$, $\frac{4}{5}$, for example? How many different fractions can they show in one string of beads? Can they do it while still making a regular pattern with colours? Children with a deeper understanding of fractions can take this to as complex a level as they wish.

> ### Answers for Pupil Book 2 page 73
>
> <u>Think and share:</u> In the 2nd bracelet, 1 out of 4 is black. $\frac{1}{4}$
>
> In the 3rd bracelet, 2 out of 5 are black. $\frac{2}{5}$
>
> In the 4th bracelet, 3 out of 6 are black. $\frac{3}{6}$
>
> Some children may notice from the picture that this is equivalent to $\frac{1}{2}$.
>
> **1** a unequal   b equal   c equal
>   d equal   e unequal   f equal
>
> **2** a equal   b unequal   c unequal
>   d equal   e equal   f unequal

Halves and quarters

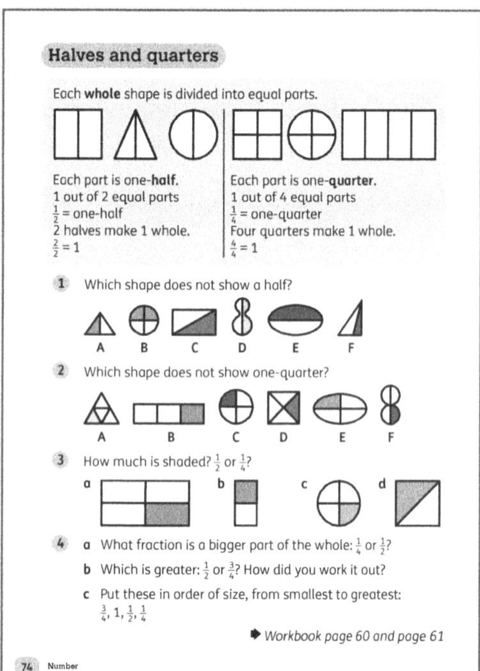

Materials

Paper plates with lines dividing them into quarters; cut-out circles (slightly smaller than the paper plates) in paper of a variety of colours, cut into quarters; glue; flashcards with the fractions $\frac{1}{2}$, $\frac{2}{2}$, $\frac{1}{4}$, $\frac{3}{4}$, $\frac{4}{4}$; flashcards with the words 'one-half', 'one-quarter', 'two-quarters', 'whole', 'three-quarters'; large sheets of paper

Warm-up

- Give the children marked paper plates, coloured paper cut into quarters and flashcards cards of the different fraction names and fractions. If necessary, remind the children of the term *three-quarters*.
- For each flashcard, the children should stick the correct number of quarters onto a plate.
- Then, as a class, work together to sort the fraction plates into groups showing the same fraction. You can stick these onto large sheets of paper to make fraction posters for the classroom wall.

Focus

- Work through questions 1–4 on **Pupil Book 2 page 74** with the class.
- Make sure you emphasise that $\frac{1}{2}$ means 1 out of 2 equal parts and that $\frac{2}{2}$ means 2 out of 2 equal parts, or 1 whole.
- The children should easily be able to see the equivalence of $\frac{2}{2}$ and $\frac{4}{4}$, and the equivalence of $\frac{1}{2}$ and $\frac{2}{4}$, although these are not the formal focus of this lesson.

Follow-up

Use **Workbook 2 page 60** to consolidate one-half and one-quarter of shapes. On **Workbook 2 page 61**, the children continue with this work but they look at how many parts of a shape are shaded.

Challenge

Ask the children to draw their own shapes or pictures divided into quarters or halves.

Dividing shapes into halves and quarters can also form the basis of a lesson in geometric art, as the children can use this as a starting point for creating their own pattern or design.

Interesting mistakes

- The children may need additional support to understand that the 2 halves or 4 quarters of a whole must be exactly the same size, particularly as the word *half* is used fairly casually in everyday life. Give them plenty of practical experience of folding and cutting shapes of different sizes to reinforce this concept.
- The children may say: 'This can't be a half because the shape is not cut into 2 parts that are the same.' Be aware that half of a shape does not have to be symmetrical. Folding a rectangle diagonally will produce 2 halves, but the fold line is not a mirror line. The halves are the same size, but they do not fit on top of each other as they would in a square.

Answers for Pupil Book 2 page 74

1. F
2. B
3. a $\frac{1}{4}$ b $\frac{1}{2}$ c $\frac{1}{4}$ d $\frac{1}{2}$
4. a $\frac{1}{2}$
   b $\frac{3}{4}$ Possible answer: $\frac{3}{4}$ is 3 parts out of 4 shaded and $\frac{1}{2}$ is 2 parts out of 4 shaded.
   c $\frac{1}{4}$, $\frac{1}{2}$, $\frac{3}{4}$, 1

Answers for Workbook 2 page 60

1. 1 part of each shape coloured
2. Possible answers (some shapes can be divided in more than one way):

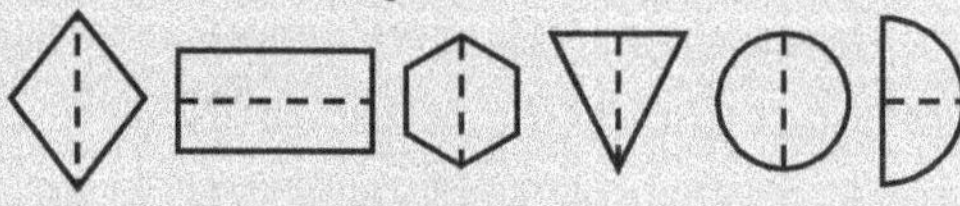

3. 1 part of each shape coloured
4. Possible answers:

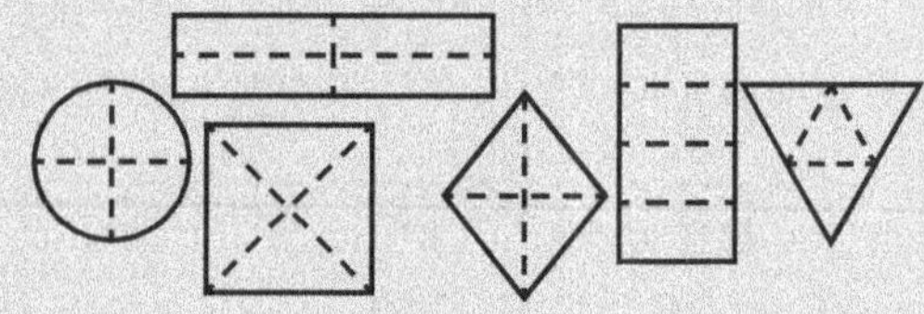

Answers for Workbook 2 page 61

1. a $\frac{1}{2}$ b $\frac{1}{4}$
   c $\frac{3}{4}$ [Provided as an example] d $\frac{2}{4}$
   e $\frac{2}{3}$ f $\frac{1}{3}$
   g $\frac{1}{4}$ h $\frac{3}{3}$

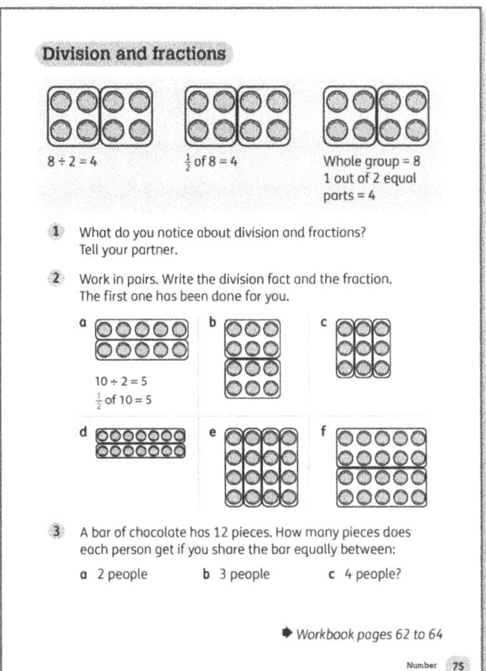

Materials

Counters, interlocking cubes (page 21) or everyday objects that can be split into fractions and put back together (such as construction blocks)

Warm-up

- Set out an array of 12 counters or objects. Ask:
  How many counters make the whole group? (12)
  How many counters make half of the group? (6)
  How can I check that it is half?
- Ask whether anyone remembers the dividing that they did in Unit 12. Set out another array of 12 counters. Ask a child to come up and check how many twos can be made from the 12 counters. (6) They can use skip counting or repeated addition to count off the twos. Ask:
  - *Who remembers how we write this as a number sentence using the division sign?* (÷)
  - *What do you notice about dividing by 2, and splitting a group in half?* (They are the same.)

Focus

- Do some more practical demonstrations of dividing by 2 and sharing a group into 2 halves. Invite the children to share their observations.
- Turn to **Pupil Book 2 page 75**. Work though the example and ask the children to talk to their partner for question 1. Then work through the first two or three parts of question 2 together.
- The children can continue with the rest of question 2 and question 3 independently, although some children may need assistance.

Follow-up

On **Workbook 2 pages 62–64**, the children continue making fractions of groups or shapes. As far as possible, let them continue this work independently.

Answers for Pupil Book 2 page 75

1 Possible answer: Dividing by 2 is the same as finding $\frac{1}{2}$ of a number. Dividing by 4 is the same as finding $\frac{1}{4}$ of a number.

2 **a** $10 \div 2 = 5$ $\frac{1}{2}$ of $10 = 5$ [Provided as an example]
b $12 \div 2 = 6$ $\frac{1}{2}$ of $12 = 6$ **c** $9 \div 3 = 3$ $\frac{1}{3}$ of $9 = 3$
d $14 \div 2 = 7$ $\frac{1}{2}$ of $14 = 7$ **e** $16 \div 4 = 4$ $\frac{1}{4}$ of $16 = 4$
f $20 \div 2 = 10$ $\frac{1}{2}$ of $20 = 10$

3 **a** 6 **b** 4 **c** 3

Answers for Workbook 2 page 62

1 **a** 3 **b** 5 **c** 4 **d** 2
e 6 **f** 1 **g** 8

Answers for Workbook 2 page 63

1 **a** $\frac{1}{4}$ of 8 is 2 **b** $\frac{1}{4}$ of 12 is 3
c $\frac{1}{4}$ of 20 is 5 **d** $\frac{1}{4}$ of 16 is 4

Answers for Workbook 2 page 64

1 1st, 2nd and 3rd shapes: 2 sections of each shape coloured; 4th shape: 4 sections coloured

2 3 cakes coloured; 2 pears coloured; 4 sweets coloured; 6 oranges coloured; 8 candles coloured

3 2 sections of each shape coloured

4 1 book coloured; 2 butterflies coloured; 4 candles coloured

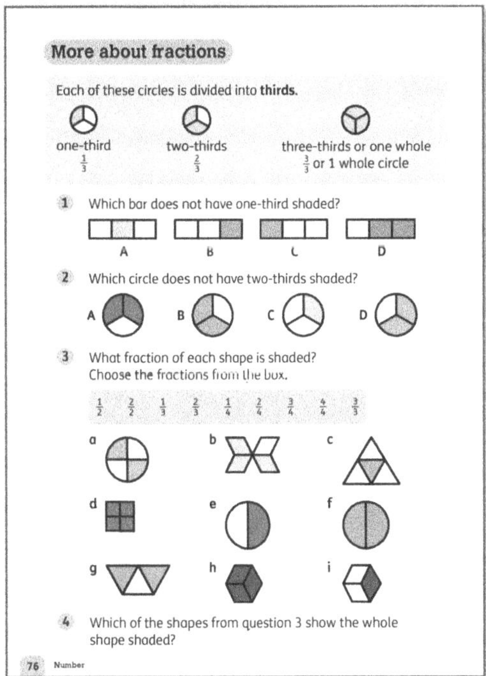

Materials

Paper plates marked in thirds; circular paper cut-outs slightly smaller than the plates, cut into thirds; glue; a selection of different shapes divided into thirds (see 'Focus' below); pattern shapes or access to an online pattern shapes tool

Warm-up

- Show the class one of the paper plates that you have marked in thirds. Explore the term *thirds* by asking:
  - *I have one whole plate. How many parts have I divided it into?* (3)
  - *Are the parts equal or unequal?* (equal)
  - *Can I call them halves? Why not?* (No, because the plate has not been divided by 2, it has been divided by 3.)
  - *Can I call them quarters? Why not?* (No, because the plate has not been divided by 4, it has been divided by 3.)
  - *Is each part larger or smaller than a half?* (smaller)
  - *Are the parts larger or smaller than a quarter?* (larger)
- Show one of the paper plates divided into quarters and halves from a previous lesson and let the children compare the sizes of the fractions.

Focus

- Introduce the terms *one-third*, *two-thirds* and *three-thirds* (or one whole).
- Show some other shapes divided into thirds. You might want to prepare drawn shapes for this, or if you have pattern shapes in your classroom, build some as models. Alternatively, use an online interactive tool. Here are some examples of shapes you might build or draw.

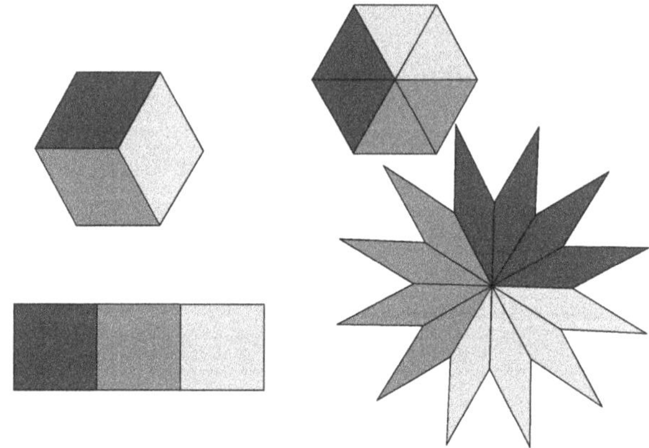

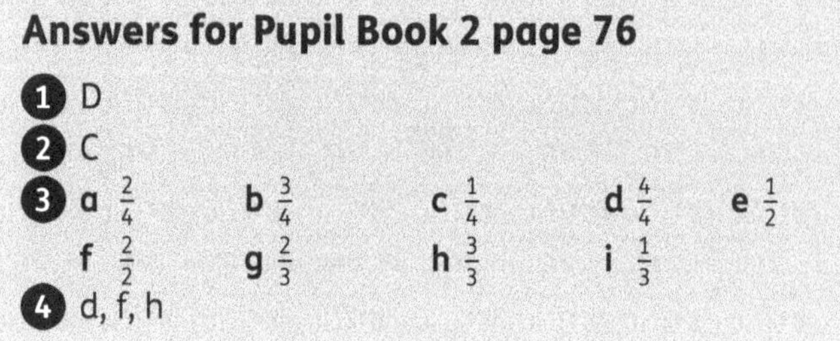

- Work through **Pupil Book 2 page 76** with the class. The children can complete questions 1–4 independently, as far as possible.

Answers for Pupil Book 2 page 76

1 D

2 C

3 a $\frac{2}{4}$ b $\frac{3}{4}$ c $\frac{1}{4}$ d $\frac{4}{4}$ e $\frac{1}{2}$
  f $\frac{2}{2}$ g $\frac{2}{3}$ h $\frac{3}{3}$ i $\frac{1}{3}$

4 d, f, h

Equivalent fractions

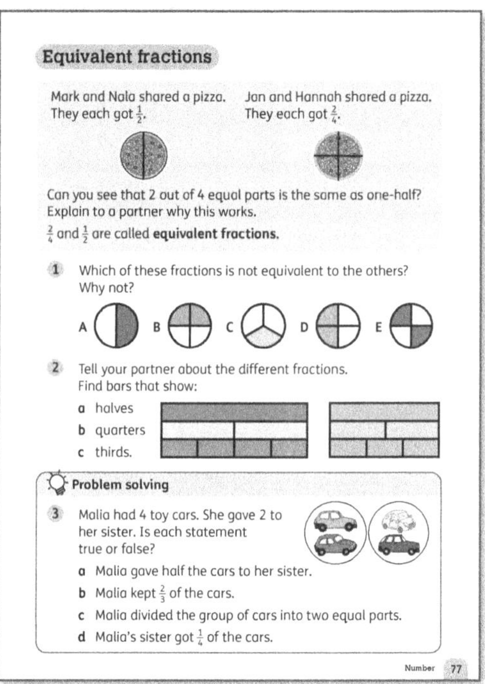

Materials

Slabs of chocolate if possible or interlocking cubes (page 21) or construction blocks

Warm-up

Introduce the idea that one fraction is *equivalent* to another fraction using practical examples.

- A fun way to introduce this is by using bars of chocolate. Open a bar of chocolate and let the children count how many blocks are in each row and how many rows there are altogether. Work out how many blocks would make half a bar. Break this bar in half. Open another bar and break it into quarters. Demonstrate that two quarters are equivalent to one half.
- If you prefer not to use food in your classroom, do similar activities using interlocking cubes.

> Whenever you work with concrete materials, encourage the class to notice what is one whole.

Focus

- Work through the explanation and question 1 on **Pupil Book 2 page 77** with the class.
- The children can complete question 2 orally in pairs.
- Problem solving: The children can do question 3 in pairs before discussing their answers as a class.

Interesting mistakes

- The children may think that $\frac{4}{4} = 4$. This is an interesting mistake that (like many errors) comes from trying to recall a 'rule'. Ask leading questions to get the children to investigate the mistake and work out the thinking behind it. Ask:
  - *Who thinks they can see why this person thought 4 is the answer?*
  - *Could it be the answer? Why not?*

- ○ *What does $\frac{4}{4}$ tell us?*
- ○ *Can someone show us in a diagram?*
- Typically, this mistake comes from seeing the $\frac{4}{4}$ as 4 = 4, rather than 4 out of 4. Once we see that the 4 at the bottom represents 4 equal parts that make up a whole, and the 4 at the top represents 4 of those 4 equal parts, it becomes more obvious that $\frac{4}{4}$ = 1.
- The children may also think that $\frac{2}{2}$ = 4. Again, encourage them to investigate the mistake. Typically, the error here is interpreting the fraction line as a sign for addition: 2 + 2 = 4. Again, focusing on what $\frac{2}{2}$ represents (using shapes or groups) will help to clarify the children's thinking.

Answers for Pupil Book 2 page 77

1 C All of the other shapes show $\frac{1}{2}$ shaded, but shape C shows $\frac{1}{3}$.

2 a left-hand diagram: middle row; right-hand diagram: middle row

  b left-hand diagram: bottom row

  c right-hand diagram: bottom row

<u>Problem solving:</u>

3 a true   b false   c true   d false

Put fractions together

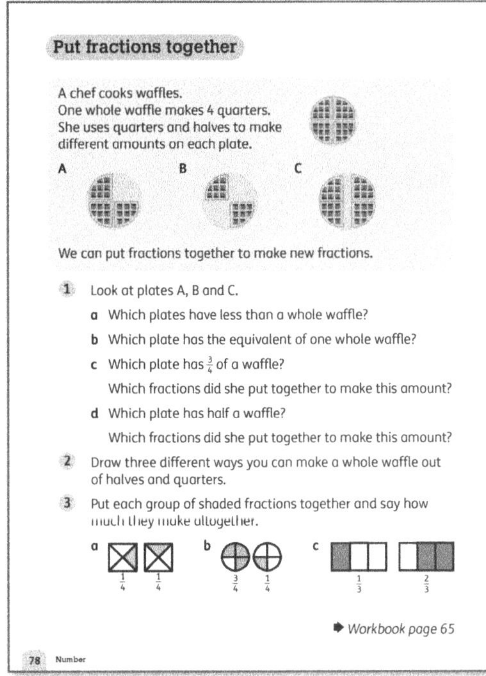

Materials

Paper plates, some whole and some cut into quarters

Warm-up

- Use whole paper plates and paper plates cut into quarters to model the waffle quarters shown at the top of **Pupil Book 2 page 78**. Make sure the children know how to represent one-quarter, two-quarters, three-quarters and four-quarters as well as one-half and two-halves.

Focus

- This lesson follows on from the previous one, as the children use their understanding of equivalent fractions in order to add and subtract.
- The children can complete questions 1–3 in pairs or independently.

Follow-up

You will need to help the children to read each question on **Workbook 2 page 65**, as there is a lot of reading here for this level. The focus should be on understanding and working out each question in a step-by-step way, rather than reading all the questions and trying to answer them all together.

Challenge

Ask the children to explain why all these numbers represent 1 whole:

$$\frac{10}{10} \qquad \frac{5}{5} \qquad \frac{2}{2}$$

Interesting mistakes

The children may say: $\frac{2}{4} + \frac{1}{2} = \frac{3}{4}$. At this stage, the children are not converting fractions numerically to the same denominator and they are not yet working with the terms numerator and denominator. Therefore, you need to make sure they picture fractions as parts of a whole. Ask:

- *How many quarters make a half?*
- *Why might someone think that we could add 2 quarters to another 2 quarters and get 3 quarters?*

Answers for Pupil Book 2 page 78

1 a A and B   b C

  c A 3 quarters   d B 2 quarters

2 Individual's answers. For example, the children might draw different arrangements of 1 half and 2 quarters; 4 quarters; 2 halves.

3 a $\frac{1}{2}$ b 1 whole   c 1 whole

Answers for Workbook 2 page 65

1 a 2   b half

  c 4   d quarter

2 a $\frac{1}{2}$ or one-half   b $\frac{2}{4}$ or two-quarters

3

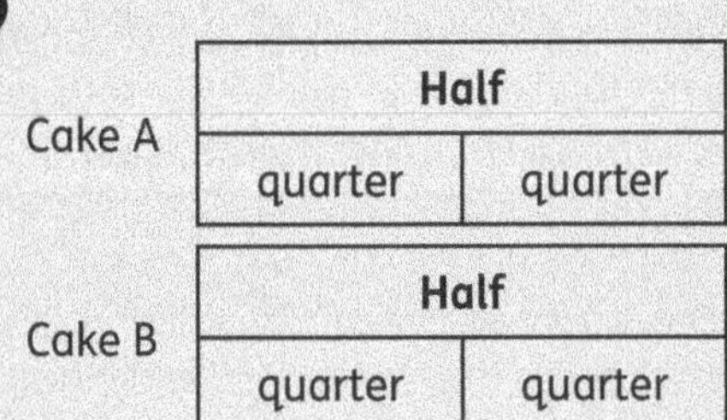

Cake A	Half	
	quarter	quarter

Cake B	Half	
	quarter	quarter

4 two-quarters

Ask questions to assess the children's understanding of fractions, for example:
* *How many parts are there in this whole* (show a shape divided into 2/3/4 parts)? (2/3/4 parts)
* *Are these parts equal?* (Show a shape divided into equal parts.) (yes)
* *Is this a half?*
* *Is this a quarter?*

* *What fraction is this* (display $\frac{1}{2}$, $\frac{1}{4}$, $\frac{3}{4}$ on cards)? (one-half/one-quarter/three-quarters)
* *How can we tell whether a shape is divided into halves?* (It is split into two equally sized shapes.)
* *How can we tell whether a shape is divided into quarters?* (It is split into four equally sized shapes.)

You could also give the children shapes with fractions shaded and ask what fraction is shaded.

UNIT 14 Time

Learning objectives
* Know the number of minutes in an hour and the number of hours in a day.
* Compare and sequence intervals of time.
* Interpret and use simple tables (in the form of calendars).

Key words
day   week   month   year   leap year
names of months   names of days   first   second
third   fourth (up to twelfth)   calendar   dates

How long does it take?

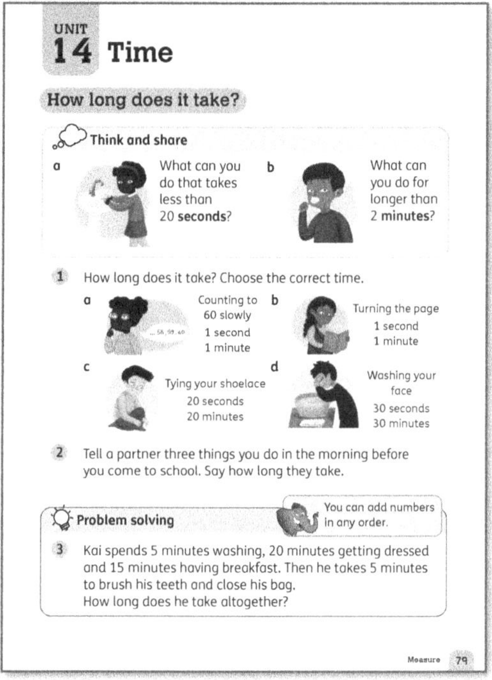

Unit introduction

Materials
Large calendar to display; flashcards with the months of the year and days of the week on them

Teaching guidance
* Revise the *names of days* of the *week* using flashcards. Display each one and ask the children to say the name. Show a day and ask which day comes before or after that day to reinforce the order.
* Use a large display calendar to teach the children the *names of the months* and the sequence in which they occur to make a *year*. The children should be familiar with how to express a *date* and be able to write the date correctly.
* Revise ordinal numbers (*first, second, third, fourth (up to twelfth)*) using days of the week and months of the *year*.

Materials
Stopwatch (an app or on a smartphone)

Warm-up
* Tell the class that when you say *Start*, they must close their eyes. When you say *Stop*, they must open them again. Say *Start*. Count to 5 slowly (about 1 second for each number), then say *Stop*. Ask: *How long did you keep your eyes closed for? Was it a long time or a short time?*
* Some children may suggest that it was 5 seconds. Write the word *second* on the board and explain that the time it took to count each number was called a second.
* Repeat the exercise, but this time they close their eyes for 10 seconds. Ask: *Was this time shorter or longer? How do you know?*

- Say: *This time, I'm not going to count, but you are going to try to guess how long it is.* Repeat the exercise, without counting, for 5 seconds (Use the stopwatch.) Ask the class to guess how long they had their eyes closed for.
- You can vary this game in many different ways, and also use it at the end of each of the upcoming lessons as a fun competition, creating a tally table of how many children got the length of time right each day.

Focus

- Ask the class to think of things you can do in 1 second. You can do this as a memory game. Start by saying: *In 1 second, I can wave goodbye* (showing the action). The next person in the circle must say: *In 1 second, I can wave goodbye and … blink* (performing each action). The next person repeats these and adds their own 1-second action. If some children run out of ideas, let the others make suggestions – examples could be sit down, turn a page, write a number, jump, say hello).
- Say: *You know how long a second takes. How long is a minute?* Let the children share what they know. Explain, if needed, that a minute is 60 seconds. You can get them to close their eyes for half a minute and ask them to guess whether it was a whole minute or half a minute. Then explain that an hour is 60 minutes.
- You can jokingly ask whether we should try closing our eyes for an hour. (This is not recommended unless you have time for a full-class nap!) Discuss different activities we do during the day and how long they take. For example, ask:
  - *What takes 5 minutes to do?*
  - *What takes 10 minutes to do?*
  - *What takes longer to do, … or … ?*
  - *What takes more than an hour?*
  - *What takes a few hours?*
- Once the children have a sense of how seconds, minutes and hours relate to each other, turn to **Pupil Book 2 page 79** with the class.
- <u>Think and share:</u> Ask the children to discuss in their groups what they can do in less than 20 seconds, and for more than 2 minutes. Compare suggestions in the class, and test them using a stopwatch.
- The children can do questions 1 and 2 in pairs.
- <u>Problem solving:</u> You can work through question 3 as a class, asking the children suggest ways of keeping track of their calculation (writing a sum, using a number line or bar model, and so on).

Challenge

What fraction of an hour is 40 minutes?

Compare times

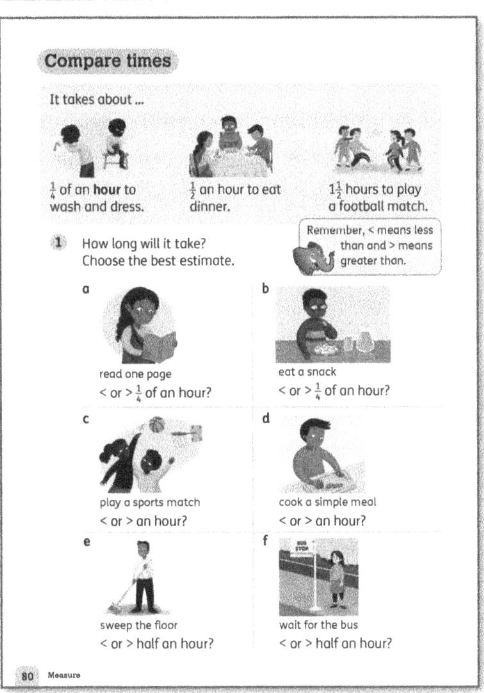

Materials

Paper plates cut into quarters (from Unit 13) objects for counting or interlocking cubes (page 21)

Warm-up

- Remind the children of the work they have done on fractions. Ask:
  - *If this paper plate is a whole hour, what part shows us half an hour?* (Half a plate.)
  - *Which part shows a quarter of an hour?* (A quarter of a plate.)
  - *How many half hours make a whole hour?* (2)
  - *How many quarter hours make a half hour?* (2)
  - *How many quarter hours are in a whole hour?* (4)
  - *Who can remember how many minutes there are in an hour?* (60)

Ask the children to work out how many minutes there are in half an hour, and how many there are in a quarter of an hour. They could use 60 objects, or interlocking cubes to model the calculation.

Focus

Discuss the pictures on **Pupil Book 2 page 80**. Before the children complete question 1, you may need to revise the < and > signs (see page 48).

Interesting mistakes

Some children may expect that there are 10 or 100 seconds in a minute, or 10 or 100 minutes in an hour, as they are not used to working with a base-60 system. When these mistakes come up, draw the children's attention to the thinking behind the mistake. For example, ask: *How many seconds are in a minute?* (60) *So, how many seconds are in half a minute?* (30) *How can we use these times to check our answers?*

Challenge

Challenge the children to work out more difficult fractions of an hour:

- *How long is $\frac{1}{3}$ of an hour?* (20 minutes)
- *How long is $\frac{1}{5}$ of an hour?* (12 minutes)
- *How long is $\frac{1}{6}$ of an hour?* (10 minutes)

Some children might like to try to solve this problem:

Two chefs are preparing meals in a competition. Chef 1 takes $\frac{1}{4}$ of an hour to prepare the ingredients and $\frac{1}{3}$ of an hour to cook the meal. Chef 2 takes 10 minutes to prepare the ingredients and $\frac{2}{3}$ of an hour to cook the meal. Who takes longer?

Answers for Pupil Book 2 page 80

1 **a** $< \frac{1}{4}$ of an hour   **b** $< \frac{1}{4}$ of an hour
    c $>$ an hour   **d** $<$ an hour
    e $<$ half an hour ($>$ half an hour if it's a very big floor)
    f $<$ half an hour ($>$ half an hour if the bus is late)

Seconds, minutes and hours

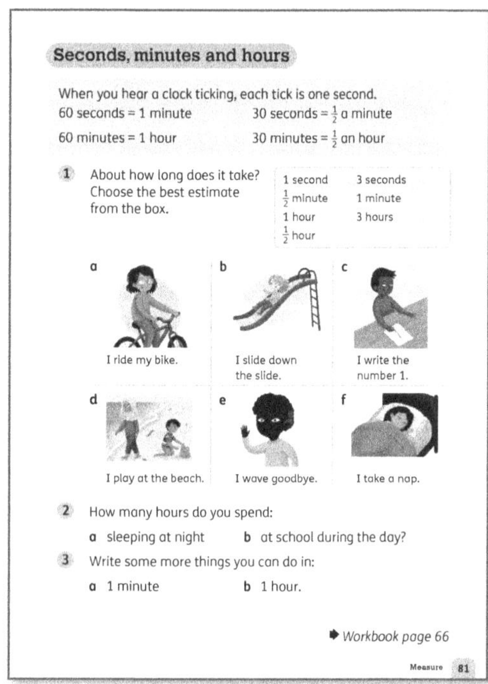

Materials

Stopwatch (an app or on a smartphone)

Warm-up

If you think it might be helpful, repeat the 'Warm-up' from the lesson 'How long does it take?' on page 110.

Focus

- Work through **Pupil Book 2 page 81**.
- For question 1, discuss with the children which answers are more reasonable for each question. They may also like to suggest other times that are not listed. For example, they might suggest 20 minutes or an hour or more for riding a bike.
- For question 2, ask questions to help the children work out how many hours they spend asleep:
  - *What time do you go to bed at night?*
  - *What time do you wake up in the morning?*
  - *How can we work out how much time passes between when you go to bed and when you wake up?*
  - *Does anyone go to bed at a different time? Is this earlier, or later?*
- Ask similar questions to help them work out how many hours they spend at school.
- For question 3, the children discuss other things they can do in a minute or in an hour.

Follow-up

For question 1 on **Workbook 2 page 66**, discuss with the children which times are less than a minute, exactly a minute and more than a minute. Complete question 1 together before moving on to question 2, letting the children work independently if possible.

Challenge

Ask the children to think of an activity to fit into each column of the tables in the Workbook.

Answers for Pupil Book 2 page 81

1 Possible answers:
    a $\frac{1}{2}$ hour/1 hour   **b** 3 seconds
    c 1 second   **d** 1 hour/3 hours
    e 3 seconds/$\frac{1}{2}$ a minute   **f** $\frac{1}{2}$ hour/1 hour

2 and **3** Individual's answers

Answers for Workbook 2 page 66

1

Less than 1 minute	Exactly 1 minute	More than 1 minute
45 seconds	60 seconds	120 seconds
1 second		2 minutes
59 seconds		1 hour
$\frac{1}{2}$ a second		1 day

2

Less than 1 hour	Exactly 1 hour	More than 1 hour
$\frac{1}{2}$ an hour	60 minutes	120 minutes
60 seconds		1 day
45 minutes		90 minutes
		62 minutes
		$\frac{1}{2}$ a day

Days and weeks

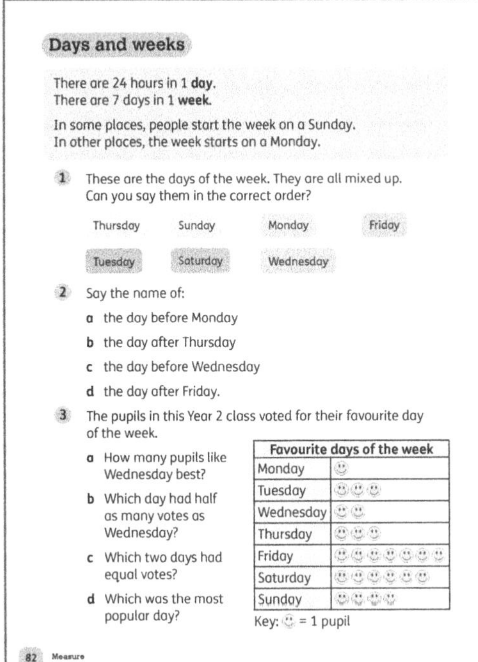

Days and weeks

There are 24 hours in 1 **day**.
There are 7 days in 1 **week**.

In some places, people start the week on a Sunday.
In other places, the week starts on a Monday.

1. These are the days of the week. They are all mixed up.
   Can you say them in the correct order?

   Thursday   Sunday   Monday   Friday
   Tuesday   Saturday   Wednesday

2. Say the name of:
   a. the day before Monday
   b. the day after Thursday
   c. the day before Wednesday
   d. the day after Friday.

3. The pupils in this Year 2 class voted for their favourite day of the week.
   a. How many pupils like Wednesday best?
   b. Which day had half as many votes as Wednesday?
   c. Which two days had equal votes?
   d. Which was the most popular day?

Favourite days of the week	
Monday	☺
Tuesday	☺☺☺
Wednesday	☺☺
Thursday	☺☺☺
Friday	☺☺☺☺☺☺
Saturday	☺☺☺☺☺
Sunday	☺☺☺☺

Key: ☺ = 1 pupil

82   Measure

Materials

A large copy of a diary page (or the simple diary page shown below drawn on the board)

Warm-up

Show the children a simple weekly diary, something like this:

Marco's weekly activity diary

Day	Activity
Sunday	swimming
Monday	free
Tuesday	piano lesson
Wednesday	tennis
Thursday	free

Ask questions about the diary, such as:

- *What is a diary? What do we use it for?* (If necessary, explain that a diary is a record of weeks and days. We can use diaries to record things that happen, or to make notes of appointments and things we need to do.)
- *What days are shown in this diary?*
- *What is the first activity of this week?*
- *Which is Marco's first free day?*
- *On how many days are there activities?*
- *Which are Marco's free days?*
- *On which days does Marco do sport?*
- *Which days are not shown on this diary?*

> You may need to adapt this activity for your class. For example, in your region, the week may start on Monday rather than Sunday.

As an additional task, the children can draw tables to show their weekly activities.

Focus

Make sure the children are familiar with the days of the week before they work through questions 1–3 on **Pupil Book 2 page 82**.

Support

In places where English is not the home language, the children may need extra practice reading and writing the names of the days of the week. Each day, make sure you write the day and date and allow the children to write it in their books.

Answers for Pupil Book 2 page 82

1. Monday, Tuesday, Wednesday, Thursday, Friday, Saturday, Sunday
2. a. Sunday   b. Friday   c. Tuesday   d. Saturday
3. a. 2   b. Monday   c. Tuesday and Thursday   d. Friday

Months and years

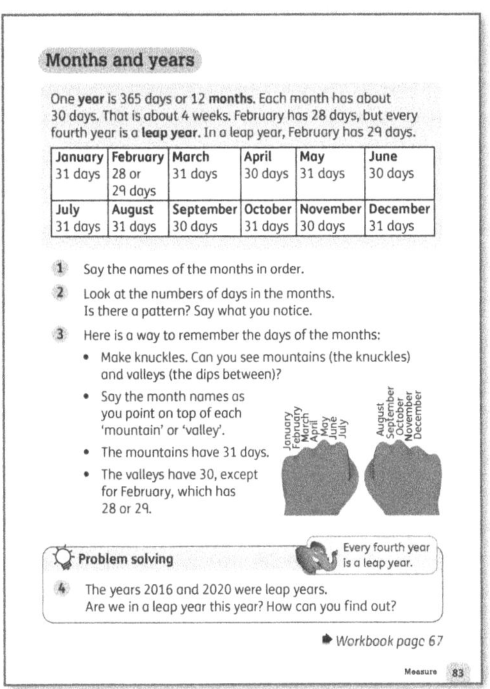

Months and years

One **year** is 365 days or 12 **months**. Each month has about 30 days. That is about 4 weeks. February has 28 days, but every fourth year is a **leap year**. In a leap year, February has 29 days.

January	February	March	April	May	June
31 days	28 or 29 days	31 days	30 days	31 days	30 days
July	August	September	October	November	December
31 days	31 days	30 days	31 days	30 days	31 days

1. Say the names of the months in order.
2. Look at the numbers of days in the months. Is there a pattern? Say what you notice.
3. Here is a way to remember the days of the months:
   - Make knuckles. Can you see mountains (the knuckles) and valleys (the dips between)?
   - Say the month names as you point on top of each 'mountain' or 'valley'.
   - The mountains have 31 days.
   - The valleys have 30, except for February, which has 28 or 29.

Problem solving

Every fourth year is a leap year.

4. The years 2016 and 2020 were leap years. Are we in a leap year this year? How can you find out?

➧ *Workbook page 67*

Measure   83

Materials

Large sheet of paper and sticky notes or stickers for birthday chart (see Warm-up)

Warm-up

- Ask the children to say which month they were born in. Use the information to create a birthday chart or pictogram for the classroom, as shown on page 114.
- Use sticky notes or stickers to place each child's name in the row of their birthday month.

Month	Birthdays
January	
February	
March	
April	
May	
June	
July	
August	
September	
October	
November	
December	

Focus

- Read the months with the children. Ask whether they know what we call the time it takes to get through all the months. Explain that there are 12 months in a year.
- Read through the information about the months on **Pupil Book 2 page 83** before completing questions 1 and 2, pointing out the term *leap year*.
- Teach the children the way of remembering the number of days in a month using their knuckles as shown in question 3.
- <u>Problem solving:</u> The children can discuss question 4 in pairs.

This might not be an appropriate activity for some children. You may wish to find an online video with a song for remembering the numbers of days in each month.

Follow-up

Use **Workbook 2 page 67** to provide practice in writing and working with the months of the year.

Answers for Pupil Book 2 page 83

1 January, February, March, April, May, June, July, August, September, October, November, December

2 Possible answers: All the months except for February have either 30 or 31 days. There is not a regular pattern. Mostly, the months alternate between one longer month and one shorter month. However, July and August both have 31 days.

3 Practical activity

<u>Problem solving:</u>

4 The answer will depend on the year it is now. (You can find leap years by counting on from 2020 in fours.)

Answers for Workbook 2 page 67

1 1st January, 2nd February, 3rd March, 4th April, 5th May, 6th June, 7th July, 8th August, 9th September, 10th October, 11th November, 12th December

2 a Ling's   b John's
   c Lean   d Individual's answer

Use a calendar

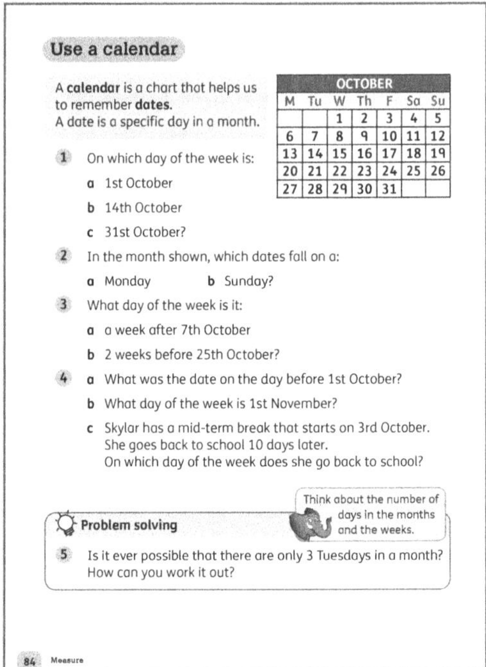

Materials

Calendars

Warm-up

Show the children some calendars. Check that they know the word *calendar* and understand how a calendar shows *dates* by asking:

- *What does this page show us?*
- *What does the heading at the top say?*
- *What do the numbers mean?*
- *How many columns are there? What do you notice at the top of each column?*
- *A month usually has 4 weeks, but in a calendar there are usually 5 rows. Why?*

Focus

- Once the children are familiar with the layout of the calendar and the abbreviations for the days of the week, they can work through questions 1–4 on **Pupil Book 2 page 84**.
- <u>Problem solving:</u> The children can share their ideas for question 5 in their groups. They could try filling in a blank calendar grid with dates to test their ideas.

Answers for Pupil Book 2 page 84

1 **a** Wednesday **b** Tuesday **c** Friday
2 **a** 6th October, 13th October, 20th October, 27th October
  b 5th October, 12th October, 19th October, 26th October
3 **a** Tuesday **b** Saturday
4 **a** 30th September
  b Saturday
  c Monday

Problem solving:

5 No. A month is always at least 28 days, which is 4 weeks, so there are at least 4 of any day.

Compare and sequence

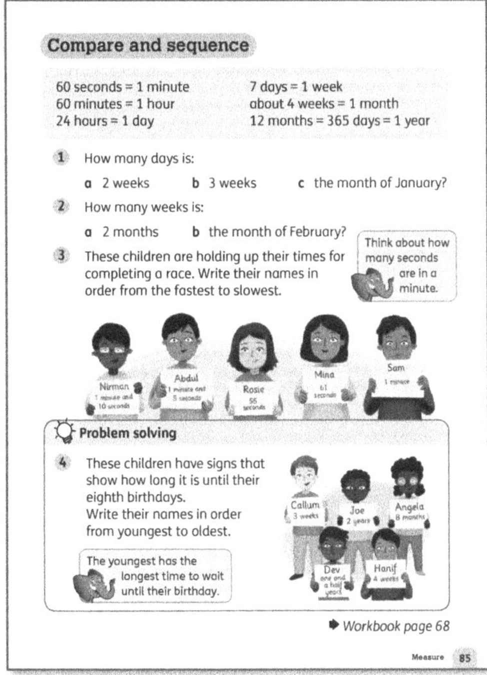

Warm-up

Discuss the different time units shown at the top of **Pupil Book 2 page 85** with the children, to check that they remember them all.

Focus

Work through questions 1–3 on **Pupil Book 2 page 85** with the class.

- Problem solving: In question 4, make sure the children understand that all the children in the question are 7, and the times shown are until their next (8th) birthday.

Follow-up

Workbook 2 page 68 provides additional practice in understanding the equivalence of different units of time and consolidates ordering and sequencing them. You could use this page for informal assessment or just for practice.

Answers for Pupil Book 2 page 85

1 **a** 14 **b** 21 **c** 31
2 **a** about 8 **b** 4
3 Rosie, Sam, Mina, Abdul, Nirman

Problem solving:

4 Joe, Dev, Angela, Hanif, Callum

Answers for Workbook 2 page 68

1 60 seconds and 1 minute; 7 days and 1 week; 2 weeks and 14 days; 1 year and 12 months; 60 minutes and 1 hour; 30 days and 1 month; $\frac{1}{2}$ a year and 6 months
2 **a** 4 weeks, 6 months, 9 months, 1 year
  b $\frac{1}{2}$ a month, 20 days, 1 month, 40 days
  c 40 seconds, 2 minutes, 3 hours, 1 day

End-of-unit check

Ask questions to assess the children's understanding of time, for example:

- *Which is longer: one minute or one second?* (one minute)
- *How many minutes are there in one hour/half an hour/ quarter of an hour?* (60; 30; 15)
- *Write these time units in order, from smallest to greatest: hour, day, year, second, week, minute, month* (second, minute, hour, day, week, month, year)
- *What is the first day of the week?* (will vary depending on country)
- *Which day comes after/before Friday?* (Saturday/ Thursday)
- *How many days are there in a week?* (7)
- *What is the first month of the year?* (January) *What is the last month of the year?* (December)
- *In which month were you born?*
- *Which month comes between June and August?* (July)
- *What is the month after March called?* (April)

Learning objectives

- Order and arrange combinations of mathematical objects in patterns and sequences.

- Conduct probability experiments with two outcomes, and present and describe the results.

Key words

pattern   possible   impossible   likely   certain
sure   maybe   outcome   experiment   results

Unit introduction

Materials
Paper; a coin

Teaching guidance
Explain to the class you are going to tell them something. They must decide whether it is something that could happen, or something that could not happen. Offer a variety of statements:
- *Next week, I am going to travel to America by bicycle.*
- *Tomorrow, it will be cloudy.*
- *Next week, I will win the lottery.*
- *A parcel will arrive for me at 3 o'clock today.*
- *If I throw this paper in the air, it will land on the floor.*
- *If I toss this coin, it will land on heads.*

Pick examples that may be entertaining for your class, such as a visit from one of their favourite pop stars or YouTube stars this afternoon, or winning a popular local TV talent contest next week.

Discuss with the class what they understand by the terms *possible* and *impossible*. Ask them to give their own examples of:
- something that is impossible
- something that is possible
- something that will definitely happen
- something that might happen, but is not definite.

As you work through the discussion, you may want to note some vocabulary on the board, such as: *possible, impossible, certain, sure, maybe, likely, unlikely.*

What is the possible outcome?

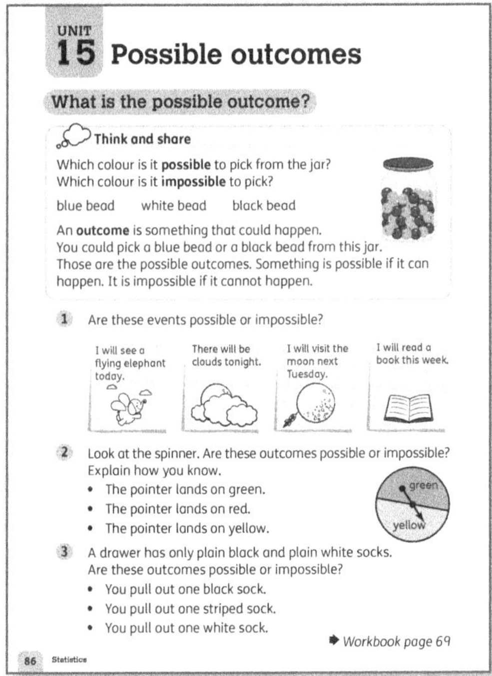

Materials
A box with objects (such as balls or counters) of two colours (or a jar with two types of object, see Warm-up below); spinner with two colours, like the one in Pupil Book question 2

Warm-up
- Show the box with balls or counters in two colours. Note that there are many different variations on offering a two-outcome scenario, such as:
  ○ a jar with two different colours of bead or stone
  ○ a box with blocks of two colours or two shapes
  ○ a bunch of straws, some plain and some striped.
- Use whatever is suitable and available for your class. Whatever combination of objects you use, make sure there are very few of one option and many of the other.
- The example below uses a box of blocks, with some blue blocks and some white blocks. Ask:
  ○ *If I take one block without looking, do you think it is more likely that I will get a blue block or a white block?*
  ○ *Is it possible that I will get a blue block? Is it possible that I will get a white block?*
  ○ *Is it possible that I will get a colour that is not blue or white? Why?*
- The children should notice that if there are only blue and white blocks in the box, the only possible outcomes are blue and white. Introduce the phrase *possible outcome* and explain that this is something that can possibly happen.

Focus

- <u>Think and share:</u> Once the children are familiar with the idea of a possible and impossible outcome, work through the questions and information on **Pupil Book 2 page 86**.
- Discuss the statements in question 1, one by one, with the whole class.
- Demonstrate how the spinner works, and ask the children to suggest possible and impossible outcomes for your spinner. Let the children work on question 2 and question 3 in pairs before discussing their ideas with the rest of the class.

Follow-up

In **Workbook 2 page 69,** the children colour the spinner in two different colours in order to create two different outcomes. If you have colour-blind children in your class, they can choose colours that they are able to distinguish, or use two different patterns, such as spots and stripes.

Challenge

Let the children think of events that fit more nuanced descriptions of probability:

- something that is possible and certain
- something that is possible, but not likely
- something that is possible and very likely, but not certain
- something that might or might not happen.

Support

Some children may need extra practice in identifying possible and impossible events. Remind them that something is impossible if it cannot happen. It is possible if it can happen.

Ask the children whether the following events are possible or impossible:

- **a** *I will see a unicorn in my garden this afternoon.*
- **b** *I will fly over the school at breaktime.*
- **c** *I will drink some water tomorrow.*
- **d** *I will do some maths today.*
- **e** *I will sleep tonight.*

Answers for Pupil Book 2 page 86

<u>Think and share:</u> It is possible to pick blue and black beads.
It is impossible to pick a white bead.
1 In order: impossible, possible, impossible, possible
2 In order: possible, impossible, possible
3 In order: possible, impossible, possible

Answers for Workbook 2 page 69

1 2 possible outcomes
(colour 1) or (colour 2)
(a colour that has not been used to shade the spinner)
2 Possible: it may happen
Impossible: it will never happen
3 Individual's answer

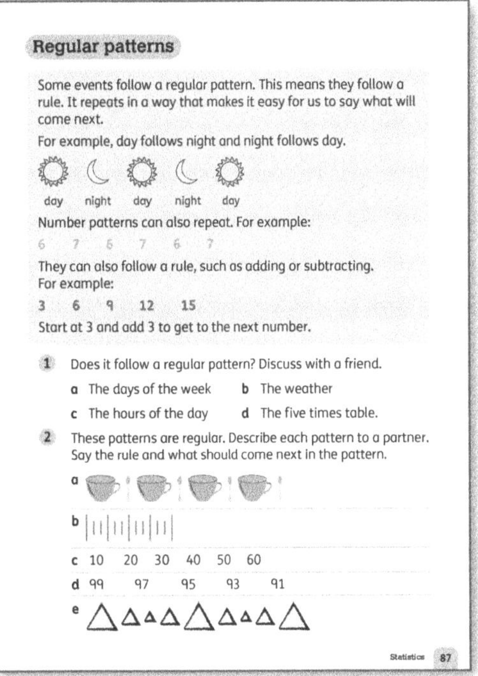

Warm-up

The children have already done *pattern* work in the context of identifying shape or number patterns. In this lesson, the focus is on regular and random patterns.

Work through the examples on **Pupil Book 2 page 87**.

Use class discussion to make sure the children understand that regular patterns are repeating patterns or patterns with a clear rule.

Focus

Either work through question 1 and question 2 with the class or let the children complete them independently.

Follow-up

The children can complete the patterns in question 1 on **Workbook 2 page 70**. Save question 2 for the next lesson, on Random patterns.

Answers for Pupil Book 2 page 87

1 **a** Yes. Every week has the same 7 days in the same order.
b No. The weather can be different every day.
c Yes. Every day has the same hours in the same order.
d Yes. You always add 5 to get to the next number.

2 **a** rule: cup, spoon; next term: cup

  b rule: long, short, short; next term: short

  c rule: add 10 to get the next number; next term: 70

  d rule: take away 2 to get the next number; next term: 89

  e rule: large, medium, small, medium; next term: medium

Answers for Workbook 2 page 70

1 **a** moon, sun, moon

  b star, cloud, cloud, star

  c eye, eye, heart, eye

  d 3, 1

Random patterns

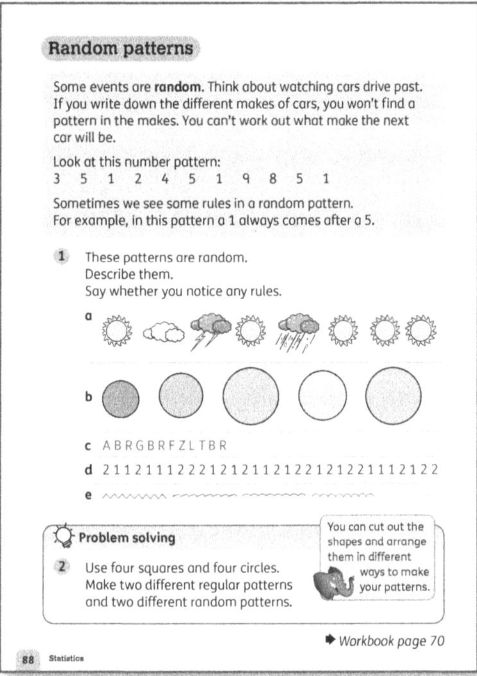

Materials

Cut-out squares and circles

Warm-up

The children have already covered regular patterns in **Pupil Book 2 page 87** and will now focus on random patterns.

Work through the examples on **Pupil Book 2 page 88**.

Use class discussion to reiterate these points:
- Regular patterns are repeating patters or patterns with a clear rule.
- Random patterns do not repeat regularly or follow a clear rule.

Focus

- Either work through question 1 on **Pupil Book 2 page 88** with the class or let them work independently.

- **Problem solving:** The children can use cut out squares and circles to create their patterns.

Follow-up

For **Workbook 2 page 70**, the children should be able to continue the patterns in question 2 in a random way. (They completed question 1 in the previous lesson.)

Answers for Pupil Book 2 page 88

1 **a** This is a random pattern of weather symbols. The only rule is that there is always a sun after a cloud, but there is no rule for how many suns there are together.

  b This random pattern is made of circles. The only rule is that the shapes are all circles. The sizes and colours of the circles are random.

  c This is a random letter pattern. The only rule is that there is an R after every B.

  d This is a random number pattern. The only rules are: the pattern uses only ones and twos; there are about the same number of ones as twos.

  e This is a random line pattern. All the lines are different. The only rule is that the pattern is made of lines.

Problem solving:

2 Individual's answers

Answers for Workbook 2 page 70

2 Individual's answer

Experiments

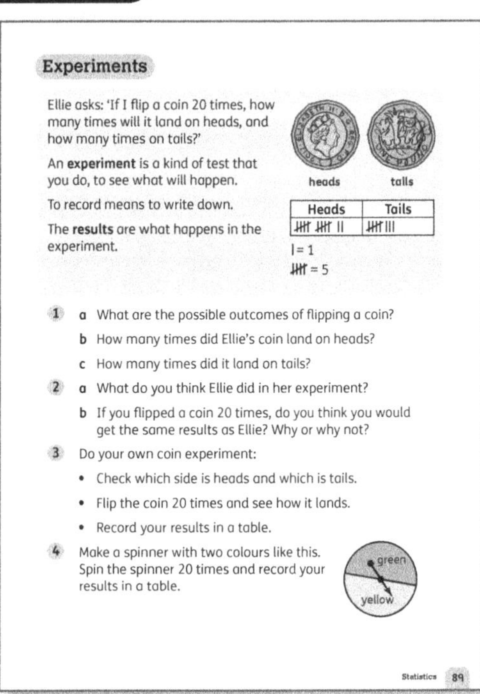

Materials

Coins for flipping; spinners with two colours (or materials for making spinners: cardboard plates, paper fasteners or wire and card for making arrow pointers)

You can easily make spinners using paper plates and paper fasteners. Make sure the pointer is attached loosely enough so that it can spin freely.

Warm-up
- Read through the information at the top of **Pupil Book 2 page 89** and discuss the terms *experiment*, *record* and *results*.

In maths, a probability experiment is a fairly simple procedure. We set up a situation that can have different outcomes, often one of two possible outcomes.
- In the event of flipping coin, the possible outcomes are: a) landing on heads, or b) landing on tails.
- In the event of a spinner with two colours, say black and white, the possible outcomes are a) landing on black, or b) landing on white.

When we do experiments, we record the actual outcomes from a given number of attempts (for example, the results of tossing a coin 10 times or 20 times or 100 times).

Choose practical experiments that are suitable for your class. Tossing a coin may be the easiest to do practically. However, if you have two-colour spinners, you can use these instead.

If using coins, you may need to explain that a coin has two sides. On many coins, one side has the head of a person. This side is usually called heads. However, sometimes the heads side has other pictures, such as an animal or a flower. The other side is called tails. It is always a good idea to look at the coin and decide which side will be called heads and which side tails, before tossing.

Focus
- Discuss question 1 and question 2 with the class.
- For question 3, the children can work in pairs to do their own coin experiment under your supervision.
- For question 4, hand out materials for making spinners. The children can choose their own colours. Discuss how they must colour the spinner to make sure the two outcomes are equally likely.

Challenge
You may want to discuss with the children how it changes the likelihood of each outcome if they vary the sizes of the two colours on the spinner. If both halves are the same colour, the outcomes are equally likely. However, if they have unequal colour areas, they change the probability.

Some children may want make spinners with three or more possible outcomes.

Answers for Pupil Book 2 page 89
1 a heads, tails   b 12   c 8

2 a Possible answer: She flipped a coin 20 times and drew a tally mark each time to show if it landed on heads or tails.
  b No. The coin will probably land differently.
3 and **4** Individual's answers

End-of-unit check

Ask questions to assess the children's understanding of probability, for example:
- *Today I will* (suggest a possible activity, for example, drive home from school). *What is the probability of that happening?* (possible/likely)
- *Tomorrow I will* (suggest an impossible activity, for example, fly to the moon). *What is the probability of that happening?* (impossible/unlikely)
- *What are the possible outcomes of flipping a coin?* (heads or tails, check for differences with local currency)
- *What experiment did you do in this unit?* (For example, the children describe the experiments they completed such as flipping a coin or using spinners.)
- *What is the rule for this sequence?* (Show regular repeating sequences of shapes or numbers, and number sequences with rules such as 'add 2' or 'subtract 3')
- Give the children a set of two shapes, and ask them to make a regular pattern, then a random pattern with them.

Mixed practice 2

Answers for Pupil Book 2 pages 90–91
1 a pineapple   b 12   c 5
  d pineapple, banana and watermelon because they are the most popular fruits
2 Individual's answers
3 a $3 \times 4 = 12$ $4 \times 3 = 12$ $12 \div 4 = 3$ $12 \div 3 = 4$
  b $2 \times 5 = 10$ $5 \times 2 = 10$ $10 \div 5 = 2$ $10 \div 2 = 5$
  c $4 \times 4 = 16$ $16 \div 4 = 4$
4 a correct because you can multiply two numbers in any order
  b correct because $\frac{1}{2}$ of a number is the same as dividing the number by 2
  c incorrect because you cannot divide in any order. The order of the numbers matters
5 a 5   b 12
6 Individual's answers to show $\frac{1}{4}$, $\frac{1}{2}$ and $\frac{2}{3}$.
7 There are **7** days in a week and **about 4** weeks in a month.
  There are **60** minutes in an hour and **30** minutes in a half hour.
8 Individual's answers
9 heads and tails
10 Individual's answers

Learning objectives

- Identify a line of symmetry on 2D shapes and patterns.

- Sketch the reflection of a 2D shape in a vertical mirror line including where the mirror line is the edge of the shape.

Key words

triangle   square   rectangle   pentagon
hexagon   sides   angles   reflection
line symmetry   symmetrical   mirror line
line of symmetry

Unit introduction

Materials

Large symmetrical shapes to display lines of symmetry (isosceles triangle, equilateral triangle, square, rectangle, pentagon, hexagon); mirrors

Teaching guidance

- Introduce the concept of *symmetry* and the word *symmetrical* by pointing out things in the classroom that have *lines of symmetry*. Explain this term. Use this activity and the next (paper folding), to revise the names of shapes: *triangle*, *square*, *rectangle*, *pentagon* and *hexagon*.

- Use a rectangular sheet of paper to demonstrate to the children that, by folding the paper in half along its length, one half fits exactly on top of the other therefore the fold is along a line of symmetry.

- Repeat this by folding the paper along its width. It is also worthwhile showing the children that folding the sheet along a diagonal produces two halves equal in size, but the fold is not along a line of symmetry as one half does not fit exactly on top of the other.

- Carry out some similar practical folding activities using the other shapes with the class. Remind them of the terms *sides*, *vertices* and *angles*.

Investigate symmetry

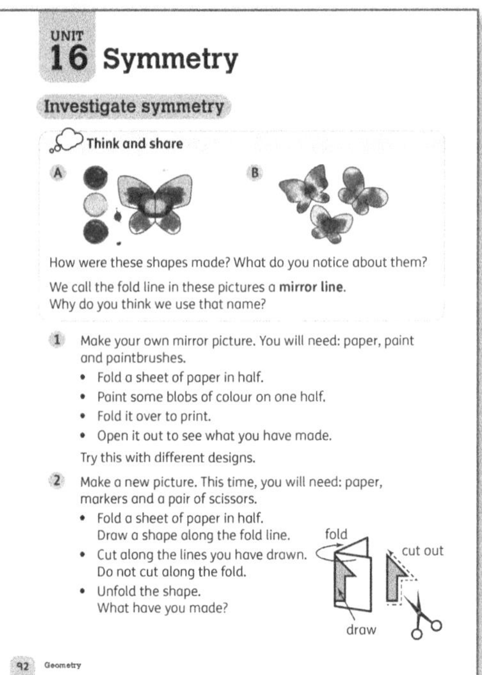

Materials

Collection of objects and pictures that show line symmetry (for example, photographs or real-life examples of symmetrical tiles, buildings, leaves, butterfly wings, fabric designs); coloured paper or card; scissors, pinboards (or dotted paper); paint; brushes; water; markers; magazines (optional); digital camera or smartphone (optional)

Warm-up

- Let the children find examples of symmetry in the environment and in nature. (As you do this, bear in mind that natural objects, such as leaves, are never completely symmetrical.) If possible, take digital photographs and use these to make a classroom display of symmetrical patterns and shapes.

- If it is not possible to take the children outside, prepare a whiteboard presentation of symmetrical objects. You will find many examples on the Internet.

Focus

- <u>Think and share:</u> Discuss the pictures on **Pupil Book 2 page 92** and talk about how the paint printings were made. Introduce the term *mirror line*.

- Follow the steps in question 1 and question 2, helping the children to make their own symmetrical printings and cut-outs.

- The children should cut shapes out of paper or card and explore whether they have any lines of symmetry by seeing if it is possible to fold one half of a shape exactly on top of the other. They could use a mirror to draw shapes with one line of symmetry.

- The children could make a scrapbook of pictures of symmetrical shapes used in designs, and patterns cut from magazines and other publications. (Ask the children to bring magazines into school prior to this activity.)

Challenge

Some children may like to do more complex symmetrical cuttings by cutting different shapes along the fold. You can also find guidance on the Internet for creating paper-doll chains.

Support

For those children who need assistance, give them step-by-step supervision when they make their paint printings. Some children may get side-tracked by trying to create their own paintings. Even if they paint on both sides of the fold, they will still create a symmetrical paint printing when they fold the paper in half and open it out again.

Answers for Pupil Book 2 page 92

<u>Think and share:</u> They were made by cutting out a butterfly shape, then painting one half of the butterfly and folding it so that the paint patterns on both halves match.

We call the fold line a mirror line because, if we fold the butterfly along the fold line and then place a mirror on the line, we can see the whole butterfly.

1 Individual's practical work

2 Individual's practical work

Line symmetry

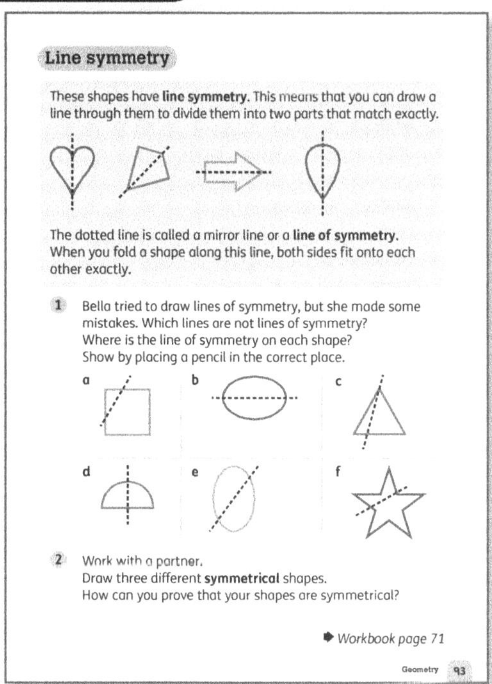

Warm-up

- Discuss the information at the top of **Pupil Book 2 page 93**. The children need to know that line symmetry means that we can find a line through the centre of a shape, and if we fold along that line, the two sides of the shape mirror each other exactly.

Focus

- Introduce the term *line symmetry* and remind the children of the meaning of mirror line. Talk about the shapes in the pictures.

- Some of the lines of symmetry in question 1 are drawn correctly and some are incorrect. Discuss with the class how we know when a line of symmetry is drawn correctly. They should notice that if the two sides are unequal, that is a good indication that the line is drawn incorrectly.
- For question 2, the children work in pairs to draw symmetrical shapes.

Follow-up

- For **Workbook 2 page 71** question 1, the children draw a line of symmetry on each shape. Note that they are only asked to draw one, but some of the shapes have multiple lines of symmetry.
- Encourage the children to compare their completed work with a partner, as some may have seen some of the lines and missed others.
- In question 2, they identify the lines of symmetry on more complex pictures. Emphasise that lines of symmetry are not always vertical.

Answers for Pupil Book 2 page 93

1 Possible answers (all these shapes have more than one line of symmetry):

a 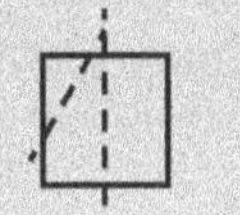 c  e 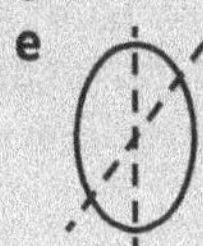 f 

2 Individual's answers. Children may prove that their shapes are symmetrical by cutting them out and folding them along the line of symmetry.

Answers for Workbook 2 page 71

1 Possible answers (some shapes have more than one line of symmetry):

2  

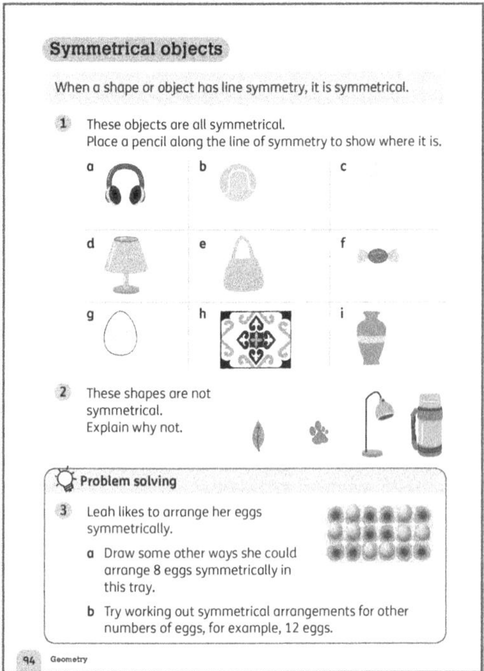

Materials

Egg boxes that hold 18 eggs (or a representation of this); balls for arranging in the egg boxes (or eggs, if you are feeling brave!); digital camera or smartphone (optional)

Warm-up

- **Pupil Book 2 page 94** continues with the work of finding lines of symmetry, but now the children look at real-life objects. Introduce the word *symmetrical* if the children have not used it before.

Focus

- Discuss the pictures in question 1. Ask questions such as:
  - *What is it?*
  - *Is it symmetrical?*
  - *Which way does the line of symmetry go – this way* (holding up a pencil vertically), *this way* (holding up a pencil horizontally) *or this way* (holding up a pencil diagonally)?
    (You might want the children to use the terms upright, sideways and slanting, as they have not yet learnt horizontal, vertical and diagonal.)
- The children then discuss why the shapes in question 2 are not symmetrical.
- Problem solving: Question 3 is a 'low floor, high ceiling' question, in which children can explore the many different ways of arranging different numbers in symmetrical patterns in a given framework. The children can explore how many different ways they can find to arrange the same number of eggs symmetrically in the box, using the egg boxes and balls.
- The children could also try arranging other numbers. This may be something you ask the children to do at home; they could take photographs and send them in to the class. You could make a display of all the different arrangements.

Challenge

Extend the egg-box activity to use a larger egg tray (such as a 30-place tray) and an unusual number such as 23.

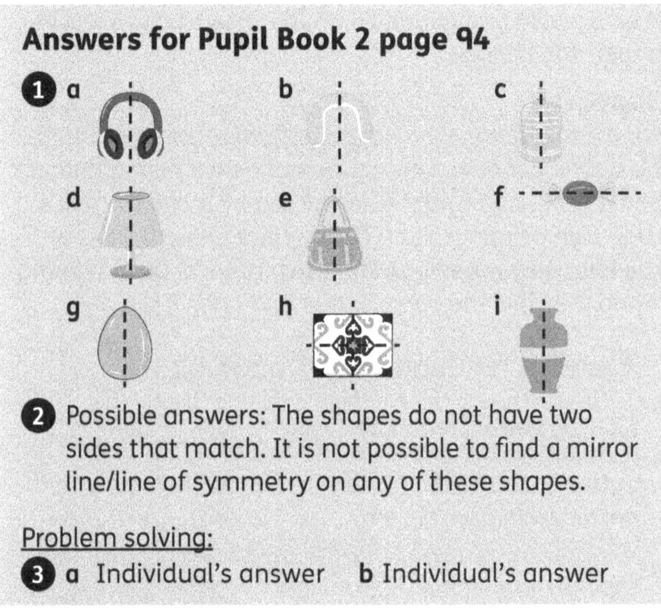

Reflections

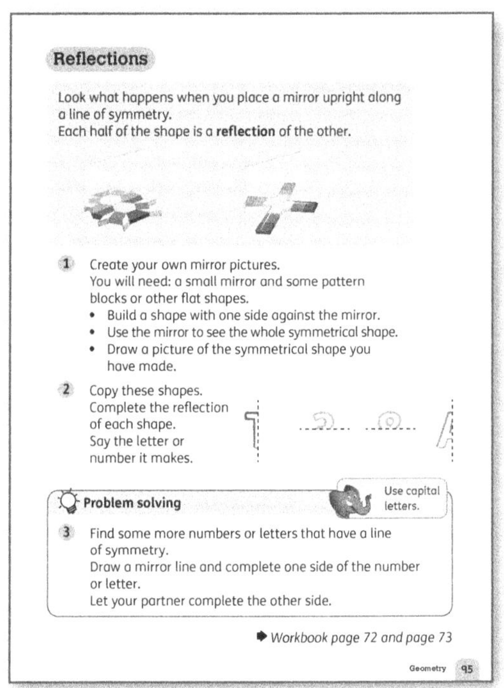

Materials

Small mirrors; pattern shapes or other blocks for constructing symmetrical shapes; squared paper

Warm-up

- Introduce the term *reflection*. A reflection is an image that we see in a mirror. In maths, we use this term to describe a shape that exactly reflects another shape along a mirror line. When we look at symmetrical shapes, the halves on each *side* of the line of symmetry are reflections of each other.
- If possible, construct shapes along one side of a mirror, as shown on **Pupil Book 2 page 95**.
- You do not have to construct the exact shapes shown in the Pupil Book – you can use simpler shapes. Start

with geometric shapes such as triangles and then build up more complex shapes.

- If you have pattern shapes in different colours, use these. If you do not have physical pattern shapes, you can use an online pattern shapes tool.

Focus

- Discuss with the children how to make the symmetrical shapes in question 1. Demonstrate how to position the mirror to get an accurate reflection. It should be upright, at right angles to the table or surface on which the shape has been built.
- Ask the children to build their own shapes and reflect them in the mirrors. They can then draw the shapes they have made.
- The children could use squared paper for question 2. If they do so, they should follow the technique of making sure that the line on each block is accurately reflected in the mirror line.
- Problem solving: For question 3, the children work in pairs to investigate other numbers and letters with line symmetry.

Follow-up

The children complete the reflections on **Workbook 2 page 72**. If you have a small mirror available, this is a fun way for the children to check their answers – their drawn object should look the same as the shape created when they hold the mirror at a right angle to the paper along the mirror line.

For **Workbook 2 page 73**, the children need to examine the patterns closely to work out where and how they are symmetrical. Encourage them to hold a pencil or ruler along the line they guess to be the line of symmetry and then to check whether all the shapes form identical reflections on both sides of the line.

Interesting mistakes

For a mathematical reflection, the mirror should be at right angles to the built shape. However, holding the mirror at different angles makes an interesting exploration. Ask questions such as:

- *Is the shape you made still symmetrical?*
- *How does it change as you change the angle of the mirror?*
- *What is happening?*
- *What happens if you move the mirror to a more closed/open position?*
- *How would you draw this on paper?*

Answers for Pupil Book 2 page 95

1 Individual's answers

2 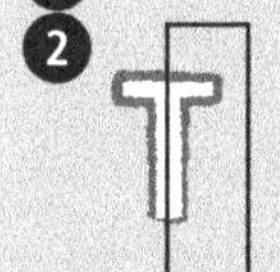 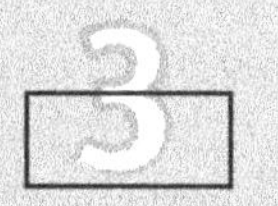  

Problem solving:

3 Possible answers: numbers: 0, 1 (if written as a vertical straight line only); letters B, C, D, E, H, I, M, O, U, V, W, X, Y

Answers for Workbook 2 page 72

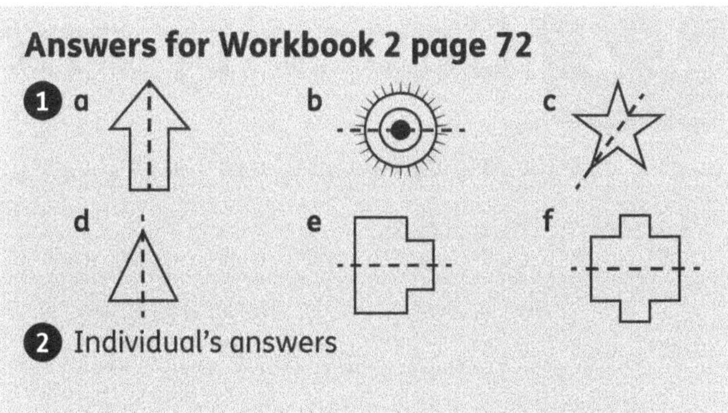

2 Individual's answers

Answers for Workbook 2 page 73

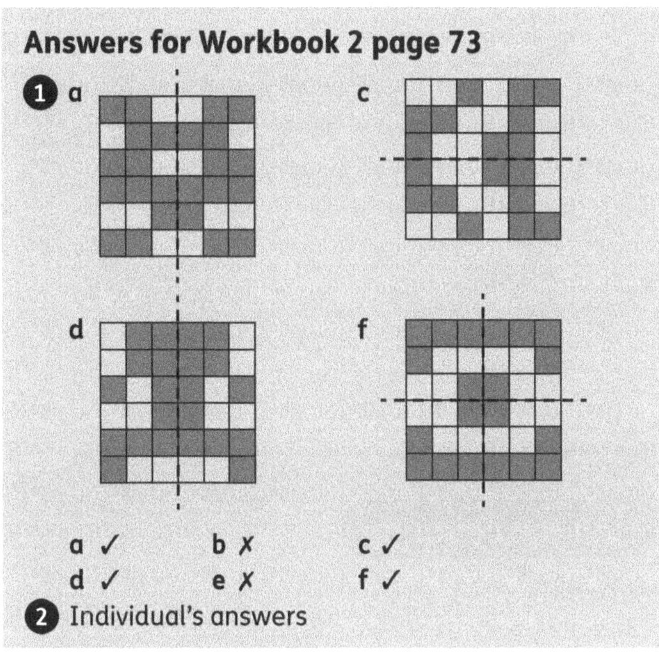

a ✓   b ✗   c ✓
d ✓   e ✗   f ✓

2 Individual's answers

End-of-unit check

Ask questions to assess the children's understanding of symmetry, for example:

- *Where is the line of symmetry down your body?* (down the middle vertically)
- *Can you name something else that has a line of symmetry?* (for example, plates, pencils, books)
- *How many ways can you fold a rectangular sheet of paper exactly onto itself?* (2)
- *How many lines of symmetry does a circle have?* (infinite)
- *Which of these capital letters are symmetrical (show different capital letters)?* (A, B, C, D, E, H, I, K, M, O, T, U, V, W, X, Y)

Give the children half shapes drawn up to a mirror line and ask them to draw in the reflection to complete the shape.

Capacity and temperature

Learning objectives

- Understand that capacity is a measure of how much liquid a container can contain or hold.
- Choose and use appropriate standard units to estimate and measure temperature (°C) and capacity (litres/ml) using thermometers and measuring vessels.
- Solve problems with addition and subtraction using quantities and measures.
- Compare and order volume/capacity and record the results using >, < and =.

Key words

capacity   container   more   less   litre (ℓ)
millilitre (ml)   pour   temperature   hot
cold   warm   cool   degrees Celsius (°C)
thermometer   measure

Unit introduction

Materials

Containers, such as drinks bottles, of capacity 1 litre, $\frac{1}{2}$ litre, $\frac{1}{4}$ litre, 2 litres, marked with their capacity in litres (or half litres and quarter litres for the smaller containers); smaller containers such as cups, beakers (no capacity marker); bowls; water or sand

Teaching guidance

The children work in groups. Give each group a set of similar *containers* of different sizes and a bowl of water or sand. The children can compare the capacities of two containers by filling one with water or sand and then pouring the sand or water into the other to see whether the second container has a smaller or greater *capacity* than the first. Encourage use of the vocabulary *container*, *pour*, *capacity*, *more* and *less*.

Compare containers

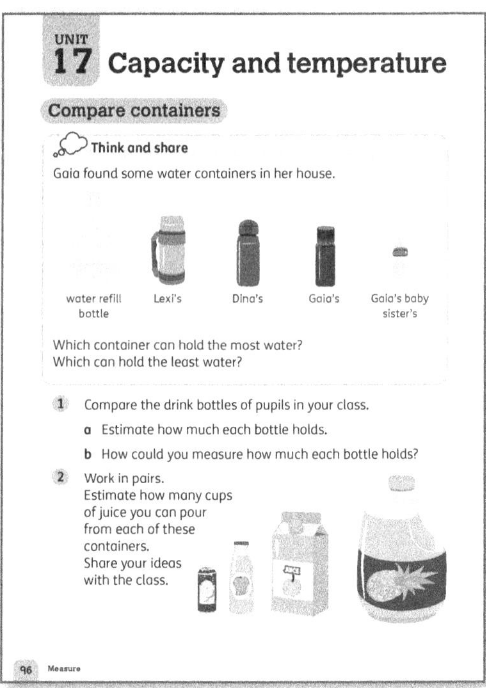

Materials

Several empty water bottles and jugs of water (if needed)

Warm-up

- Think and share: Discuss the pictures of the different water containers on **Pupil Book 2 page 96**. Ask:
  - *What can you see in the pictures?*
  - *What do we use these things for?*
  - *What is the same and what is different about these containers?*
  - *Which containers do you think would be useful for bringing to school?*
  - *Why would the big container be too big to bring to school?*
- Ask the children to take out their own juice or water bottles. Discuss why these are good sizes for bringing to school.

Focus

- Work through question 1 and question 2.

Support

- Continue the practical work from the unit introduction if any children need more support in comparing which of two containers holds more. Clear plastic or glass jugs work best for this as they allow the children to see the level of the water through the container.
- Fill a larger jug. As you pour, ask the children to say when it is half full, nearly full and completely full.
- Show a smaller, empty jug. Ask: *Do you think this one will hold more or less? Let's check.* Again, as you *pour*, let them note when it is half full, nearly full and completely full.

- Point out that you have completely filled the second jug, but the other one is not empty.
- You can repeat this several times (pouring from smaller to larger jug, as well as from larger to smaller) to compare different containers.

Interesting mistakes
- The children have already worked with mass and may initially think in terms of how heavy each bottle would be. The largest bottle would in fact be far too heavy to carry, say, to school.
- Distinguish clearly between mass (how much something weighs) and capacity (how much something can hold).
- Although a bottle with a larger capacity would be heavier when it is full, it could be lighter than one with smaller capacity when it is empty.
- Also discuss the idea that a 1-litre bottle has a capacity of 1 litre, even when it is not full.

Answers for Pupil Book 2 page 96

<u>Think and share:</u> The water refill bottle can hold the most water.
Gaia's baby sister's bottle can hold the least water.
1 and **2** Individual's answers

Units of capacity

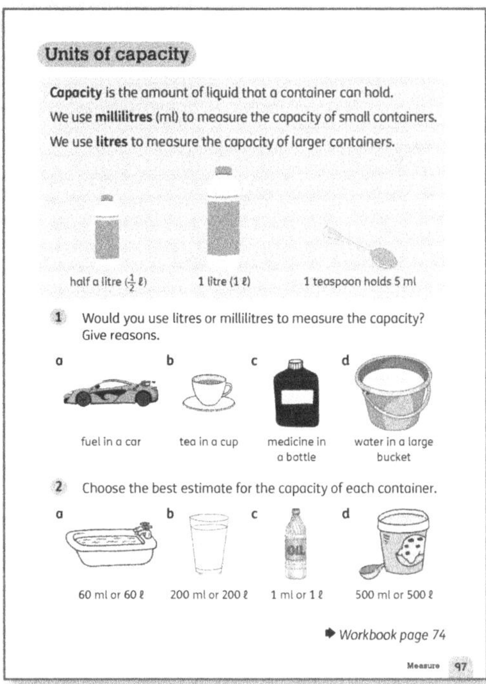

Materials
Measuring spoons and cups marked in millilitres; bottles marked in litres

Warm-up
- Once you are certain that the children understand the concept of capacity as a *measure* of how much a container holds, move on to the formal units for capacity.
- Show the measuring spoons and cups, and let the children explore the markings. Discuss what they think these markings mean.

Focus
- Discuss the information at the top of **Pupil Book 2 page 97** and introduce the terms *millilitre* and *litre*. Explain that we use millilitres for small amounts, such as teaspoonfuls and that a litre is a larger amount. Let the children discuss where they have seen these measurements. Draw the children's attention to the abbreviations for millilitre and litre, that is *ml* and *ℓ*.
- Ask the children to discuss question 1 and question 2 with a partner and then share their ideas with the class.

Follow-up
For **Workbook 2 page 74**, discuss the sizes of the different containers with the children, and how we write each measurement. You may choose to work through this page with the class or let them complete it independently.

Challenge
Ask the children whether there are any other words they have heard that these units remind them of. Challenge the children to find other units that use 'milli-', and to work out what this means. They may notice the similarity with the terms millimetre and metre. You can explain that there are 1000 millilitres in 1 litre and 1000 millimetres in 1 metre.

Note that the children will only deal formally with thousands next year, so this is extension material at this stage.

Answers for Pupil Book 2 page 97
1 a litres; large container
   b millilitres; small container
   c millilitres; small container
   d litres; large container
2 a 60 ℓ   b 200 ml   c 1 ℓ   d 500 ml

Answers for Workbook 2 page 74
1 a about 2 litres   b about 5 litres
   c about 10 litres   d about 100 litres
   e about 500 litres   f about ½ litre

Working with capacity

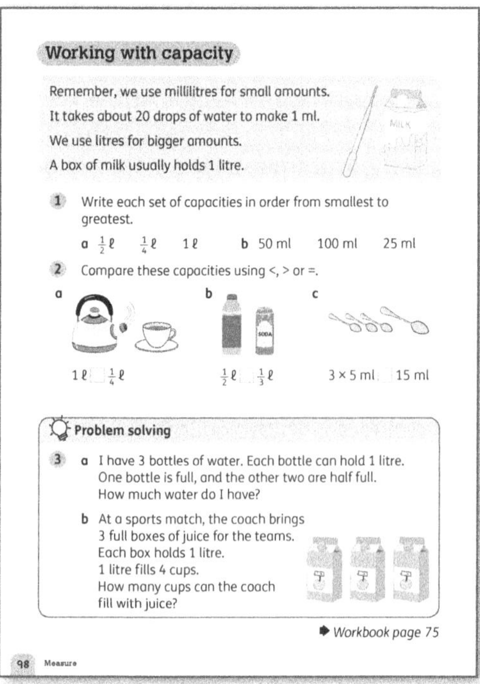

Materials

Droppers (if available); teaspoons and larger spoons; cups; water

Warm-up

Let the children do some practical activities to find out how many drops of water fill a teaspoon, how many teaspoons fill a larger spoon or cup, and so on.

> At this stage, because the children have not yet worked with thousands, avoid presenting litres as 1000 ml. If you feel that some of the children are comfortable with this number range, you can introduce it. As the children have worked with fractions, you can talk about half litres and quarter litres.

Focus

- Work through question 1 and question 2 on **Pupil Book 2 page 98** with the children.
- Problem solving: You may wish to present question 3 using a number talk (see 'Number talks', pages 17–18). Alternatively, present the parts of this question and then let the children work in pairs or groups to decide how to find the answers.

Follow-up

The children can complete **Workbook 2 page 75** independently. Check the answers together as a class.

Support

- You may need to revise the < and > signs, and discuss the procedure for ordering numbers from the smallest to the greatest.

- Ask the children to recall how we compare numbers. Ask: *How do we decide which number is smaller and which is greater?* Check that the children remember how to look at the tens and ones to decide.
- Revise: 50 ml is the same as 5×10 ml; 100 ml is the same as 10×10 ml, and so on.

Answers for Pupil Book 2 page 98

1 a $\frac{1}{4}\ell, \frac{1}{2}\ell, 1\ell$ b 25 ml, 50 ml, 100 ml
2 a $1\ell > \frac{1}{4}\ell$ b $\frac{1}{2}\ell > \frac{1}{3}\ell$ c 3×5 ml $= 15$ ml
3 Problem solving:
  a 2 litres   b 12 cups

Answers for Workbook 2 page 75

1 Possible answers (as sizes of the items may vary):
  a less than 1 litre   b more than 1 litre
  c more than 1 litre   d less than 1 litre
  e about 1 litre   f more than 1 litre

Temperature

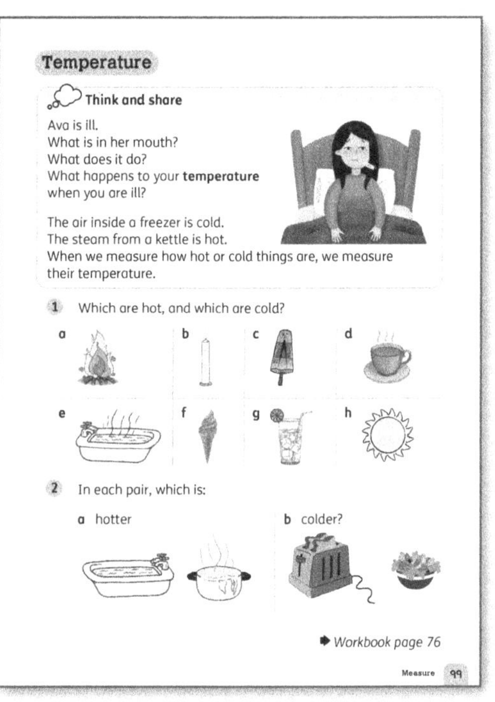

Materials

Ice; cold water; hot water in a flask; cups

Warm-up

Before we look at *temperature* in degrees and measure temperature on thermometers, it is useful for the children to think about their experience of *warm*, *hot*, *cool* and *cold* things.

- Prepare some cups of water at different temperatures. Take care not to bring boiling water to class as it is a safety hazard for young children.
- Set out cups of ice water, room-temperature water and warm or hot water. Let the children feel the different cups of water and discuss which is the warmest and which is the coolest.

- Ask them to think of other things that are very cold, very hot or warm.

Focus

- <u>Think and share:</u> Discuss what the children see in the picture of Ava on **Pupil Book 2 page 99**. Talk about the questions.
- For question 1 and question 2, the children discuss the hotter and colder items in the pictures.
- Encourage the children to think of other things at home that are very hot or very cold. If necessary, give prompts such as:
  - *What do we use in the kitchen that gets very hot? Are there any things you aren't allowed to use on your own?*
  - *What do we have to do to eggs before we eat them?*
  - *What about cold places in the kitchen – where are they?*
  - *Which is colder – the fridge or the freezer?*
  - *Why does a freezer have to be colder?*
  - *What kinds of things go in the freezer?*
  - *What happens if you take things out of the freezer and leave them on the counter?*
  - *What about things that are neither hot nor cold to touch? Which feels cooler – a pencil or a stapler? Does the wall or the desk feel cooler?*

Follow-up

The children complete **Workbook 2 page 76** independently. Help with reading the instructions if necessary.

Answers for Pupil Book 2 page 99

<u>Think and share:</u> A thermometer is in Ava's mouth. It measures her temperature.
When you are ill and have a fever, your temperature increases.

1 a hot   b hot   c cold   d hot
    e hot   f cold   g cold   h hot

2 a the pan   b the toast

Answers for Workbook 2 page 76

1 a cool   b cold   c cold   d hot
    e warm or hot   f cool or cold

2 a the sun   b a fire

3 Individual's answers, for example an ice-cream or a bottle of juice

Measure temperature

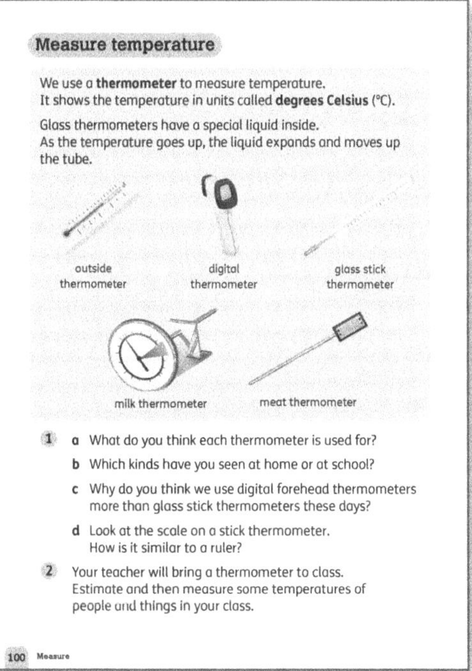

Materials

A range of different types of thermometer

Until quite recently, the most common type of oral thermometers was a glass tube with liquid inside. Digital thermometers are now far more common and the children are more likely to see temperatures as a number on an LCD screen on a handheld or even digital oral thermometer. Nonetheless, it is useful to show the children either pictures or real examples of glass (liquid) thermometers, as it is a clear visual aid to see the liquid rising higher for a higher temperature and falling lower for a lower temperature.

Try to give the children experience of seeing temperatures measured in several different ways:
- as a rising line from the bulb in a glass thermometer
- as hotter or colder along a horizontal number line
- as a number on a digital or LCD display.

Many schools have handheld thermometers for taking the children's temperatures. If possible, have a school thermometer available, as well as any others you can source.

Warm-up

- Ask the children whether they know know the word *thermometer*. Let them examine a range of thermometers. Discuss how they work and how we read the temperature shown on each thermometer.
- If you do not have a range of thermometers available, you can use the pictures on **Pupil Book 2 page 100**.

Focus

* Read the information on **Pupil Book page 100**, introducing term *degrees Celsius (°C)*. Work through question 1 and question 2 with the children.

Answers for Pupil Book 2 page 100

1 a outside thermometer: measures the air temperature

digital thermometer: measures a temperature from the skin; can also be used to measure the temperature of any external surface

glass stick thermometer: measures a person's temperature (e.g. in the mouth or under the arm)

milk thermometer: measures the temperature of liquid in a pot or jug (also known as a candy thermometer)

meat thermometer: measures the temperature inside a solid piece of food such as meat or dough, to work out whether it is cooked

b Individual's answers

c They are quicker and easier to read. We do not need to wash and sterilise the thermometer between people. Glass stick thermometers take longer to read. They are also safer because glass thermometers break easily and broken glass is dangerous.

d It has marks at equal intervals with smaller marks between. There is a value at each mark.

2 Practical work using a thermometer

Ask questions and carry out practical activities to assess the children's understanding of capacity, for example:

* *What does ℓ stand for?* (litres)
* *Arrange these containers in order of capacity, from smallest to greatest.* (Give the children some containers of different sizes.) *Show me how you decided.* (For example, the child demonstrates that the size of one container is larger than another or pours water or sand to show which container can hold the most/least).
* *How many of these bottles are needed to fill a 1-litre bottle?* (Show a small bottle, such as a 250-ml bottle, and a 1-litre bottle.) (4)
* *How many 1-litre bottles are needed to fill this container?* (Show a 1-litre bottle and a much larger container, such as a bucket.) *Show me how you know.* (For example, the child demonstrates filling the 1-litre bottle with sand or water and measuring how many will fill the larger container.)
* *What do we mean by temperature?* (Temperature is a measure of heat.)
* *How do we measure a temperature?* (thermometer).
* *Which is hotter: 30 degrees or 40 degrees?* (40 degrees)
* *What is your body temperature when you are healthy?* (range of 36 degrees to 37 degrees)
* *What happens to your temperature when you are ill?* (If you are ill and have a fever your temperature goes up.)

UNIT 18 More about time

Learning objectives

* Read and record time to five minutes in digital notation (12-hour) and on analogue clocks.

* Tell and write the time to five minutes, including quarter past/to the hour and draw the hands on a clock face to show these times.

* Know the number of minutes in an hour and the number of hours in a day.

Key words

clock   minute   hour   quarter hour
quarter to/past   digital

Materials
Analogue and digital clocks

Teaching guidance
By now, the children should know that seconds are shorter than *minutes* and that minutes are shorter than *hours*. They should also realise that there are 60 seconds in 1 minute and 60 minutes in 1 hour and 24 hours in a day. Spend a bit of time revising these terms and concepts.

* Revise how long a second is. For example, it takes about a second to say 'one hippopotamus'.

- Use the terms *clock* and *digital* and display a digital clock, or allow the children to observe their own watches or phones to see what happens in 1 minute. If possible, show them on an analogue clock that the second hand of a clock makes 60 small moves around the clock face in 1 minute, so one tick of the clock is heard each second.
- Similarly, the seconds display of a digital clock or timer will count to 60 in 1 minute. Use this to remind the class that 60 seconds is equivalent to 1 minute and that 60 minutes is equivalent to 1 hour.

Time on the clock

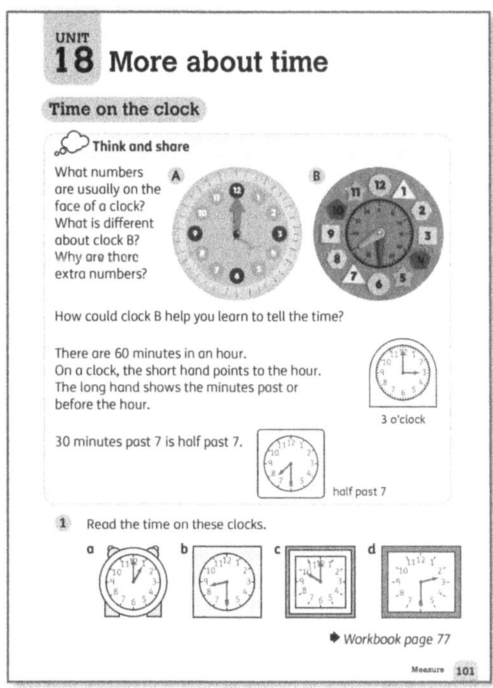

Materials
Materials for making clock faces (paper plates and large tin lids to use as templates; card; paper fasteners)

Warm-up
If the children have not done so before, have them make large clock faces (20 cm in diameter) using a paper plate or a tin lid as a template for marking and cutting out a cardboard disk, and two strips of cardboard for the hands. They can use a paper fastener to hold the hands in place.

They could also make a digital clock by writing numbers on cards and arranging the cards in a suitable order so, for example, 10 o'clock would be represented by 10:00.

Focus
- The children work in pairs. One child says a time (half past or on the hour) and the other child has to move the hands of a clock face to show this. Alternatively, ask one child to set the time on a clock face and the other child has to say the time.
- In their pairs, the children use a digital clock or a cardboard model of a digital clock with spaces for the display. One child says the time and the other sets the display using a 12-hour display option.

- The children should already be familiar with the time on the hour, but it is worth spending some time reviewing this using both an analogue clock and a digital clock.
- Show the children the positions of the hands on an analogue clock at 1 o'clock, 2 o'clock, and so on to 12 o'clock.
- Repeat this with a digital clock. Demonstrate that, starting from 12, it takes exactly one hour for the minute (long) hand of a clock to move once around.
- The children should already be familiar with the fraction 'one-half', so you can use this to introduce the idea that halfway around is 'half past the hour'.
- Think and share: Turn to **Pupil Book 2 page 101** and discuss the pictures and information with the class.
- You can also complete question 1 with the class.

Follow-up
For **Workbook 2 page 77**, the children need to look at each picture and say what it shows. They suggest a time when that activity or event might take place and draw the hands on the clock. They can stick to times on the hour and half hour, or some children may already be able to fill in more specific times.

Answers for Pupil Book 2 page 101

Think and share: The numbers that are usually on the face of a clock are the numbers 1–12.
Clock B also has numbers for the minutes past the hour, in five-minute intervals. These tell us the number of minutes past the hour.
Clock B helps us learn to tell the time because it tells us the number of minutes that the long hand is pointing to.

1 a 1 o'clock   b half past 8
    c 10 o'clock   d half past 2

Answers for Workbook 2 page 77

1 Possible answers are drawn clock hands and symbols to show, for example:
   a day, 7 o'clock (provided as an example)
   b day, half past 10   c day, 11 o'clock
   d night, 7 o'clock   e day, half past 3
   f night, half past 6   g night, 10 o'clock
   h day, 8 o'clock

Quarter hours

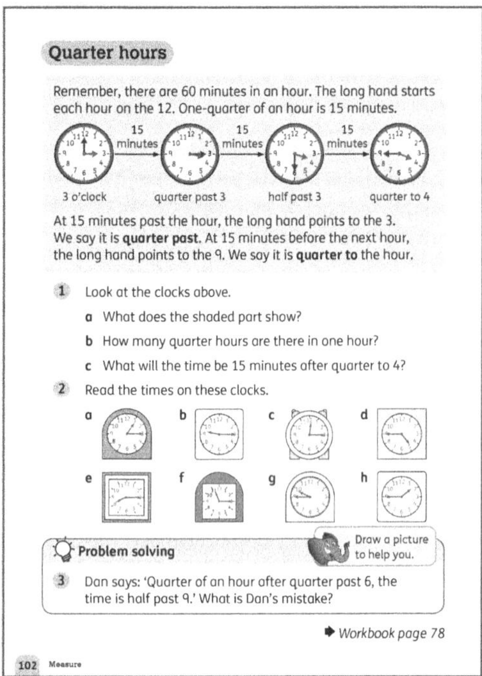

Materials
Clock faces made in the previous lesson; analogue clocks with moveable hands

Warm-up
- Show a clock face to the class. Revise the concept of an hour as 60 minutes and invite the children to demonstrate how far the minute hand moves in one hour and in half an hour.

Focus
- Then introduce the concept of a quarter of an hour, including the terms *quarter to* and *quarter past* an hour. Use the clock faces and information on **Pupil Book 2 page 102** to demonstrate and talk about times at quarter past and quarter to the hour.
- Work through question 1 and question 2 with the class.
- Problem solving: The children can use an analogue clock with moveable hands to explore question 3.

Follow-up
For **Workbook 2 page 78**, the children complete the times on the clocks independently. You may need to do some examples on the board to get them started.

Challenge
You can offer problem-solving questions that use quarter hours and half hours, for example:

- *Bo likes watching online videos that are 15 minutes long. He has 2 hours to watch videos, but he needs to take a 20-minute break in that time, too. How many videos can he watch?*
- *Eva likes to watch a programme on a subscription streaming service. Each episode is half an hour long. She is allowed to watch TV in the evening from 6 o'clock to 7.15. How many episodes can she watch?*

How much extra time would she need to fit in another episode?

Answers for Pupil Book 2 page 102
1 **a** quarter of an hour or 15 minutes
   b 4   **c** 4 o'clock
2 **a** quarter past 1   **b** quarter past 9
   c quarter past 12   **d** quarter to 5
   e quarter past 8   **f** quarter past 11
   g quarter to 10   **h** quarter to 2

Problem solving:
3 He has moved both hands on the clock round one quarter, instead of just the hour hand.

Answers for Workbook 2 page 78
1 **a** quarter past 11 (Provided as an example)
   b quarter past 7   **c** quarter to 2
   d quarter to 6   **e** quarter past 12
   f quarter past 9   **g** quarter to 3
   h quarter to 8

Digital clocks

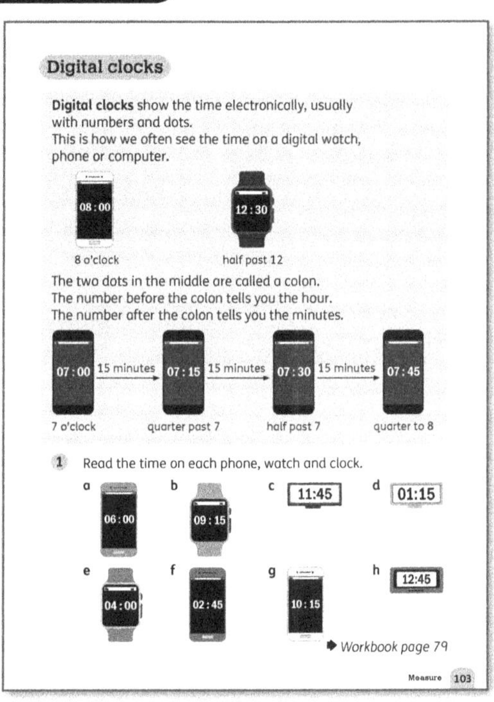

Materials
Digital watch or clock; cardboard model of a digital clock

Warm-up
- Demonstrate how to show the time on a digital clock. The children need to know the following:
  - The number before the dots tells us the hour.
  - The number after the dots tells us the minutes.
- Remind the children that there are 60 minutes in an hour and explain that :30 means half past. They also need to know that :15 is quarter past and :45 is quarter to.

- Give them plenty of time to practise writing digital times on the board or using a model to show them.

Focus
- Work through **Pupil Book 2 page 103** together.
- Explain that at quarter past 12, the time reads 12:15 and at half past 12, the time reads 12:30. Describe how the minutes continue to go up by 1 minute until we reach 59 minutes past 12 (12:59) and one minute after this, the time goes to 1 o'clock. One full hour after 12 o'clock is 1 o'clock. Demonstrate this on a model of a digital clock.

Follow-up
For **Workbook 2 page 79**, the children read the times on the clocks and write the time under each one, using words and numbers such as '11 o'clock' or 'half past 11'. Do some examples on the board before asking them to complete the rest independently.

Answers for Pupil Book 2 page 103
1 **a** 6 o'clock **b** quarter past 9
c quarter to 12 **d** quarter past 1
e 4 o'clock **f** quarter to 3
g quarter past 10 **h** quarter to 1

Answers for Workbook 2 page 79
1 **a** 11 o'clock **b** 6 o'clock
c 7 o'clock **d** half past 10
e half past 3 **f** half past 5

More about clocks

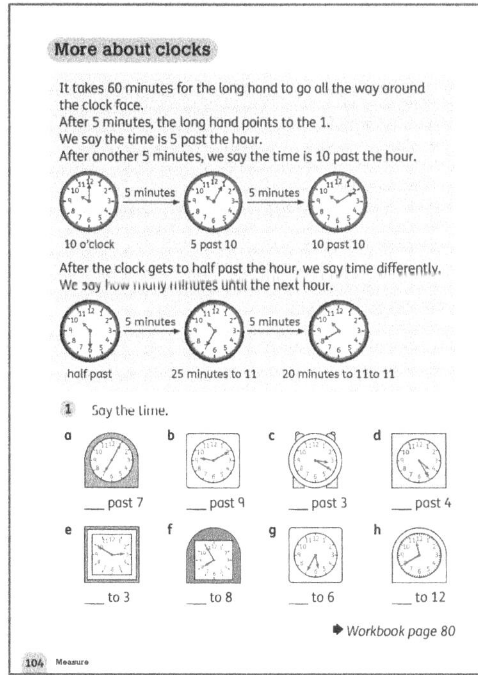

Materials
Analogue clock with moveable hands

Warm-up
- Use the information on **Pupil Book 2 page 104** and a clock with moveable hands to demonstrate times at 5-minute intervals. Start by showing a time on the clock, such as 10 o'clock. Then ask: *What will the time be 5 minutes after 10 o'clock?* If they are not sure, remind them that we already know that 15 minutes after 10 o'clock will be quarter past 10. Ask: *How many 5-minute bits make 15 minutes?* (3)
- Show that we can skip count in fives all the way around the clock: 5 minutes, 10 minutes, 15 minutes (quarter past), 20 minutes, 25 minutes, 30 minutes (half past), and so on. Repeat, counting with the class in fives, up to 60.
- Explain that this shows us that each time the minute hand gets to the new number, 5 minutes have passed. Demonstrate how we name these different times: *5 past, 10 past, 20 past* and *25 past*.
- Give the children some opportunities to name the times between a given hour on the clock and half past that hour. Make sure they are able to do this.
- Then introduce the times between half past and the next hour (*25 to, 20 to, quarter to, 10 to* and *5 to*).

Focus
- Work with the children through question 1 on **Pupil book 2 page 104** and allow them plenty of practice using the clock with moveable hands.
- Ask the children to show 5 o'clock on the clock with moveable hands. Then ask them to show:
  - the time for every 5 minutes past the hour, from 5 o'clock to half past 5
  - the time for every 5 minutes after half past 5 until they reach 6 o'clock.

Follow-up
The children complete **Workbook 2 page 80** independently to consolidate their knowledge of how to write analogue times.

Support
- If some children are finding the times in this lesson difficult, spend some time revising *quarter past* and *quarter to*.
- Remind the children that a quarter of an hour is 15 minutes. To get to 15 minutes past the hour, the minute hand has to move 15 times, once for each minute.
- Demonstrate the movement it makes in a 5-minute interval. Focus on the 5-minute intervals between, say, 12 o'clock and quarter past, until the children are comfortable with this interval.
- Do the same for quarter to 1, and the times between quarter to 1 and 1 o'clock.
- Also, from time to time throughout the lesson, ask the children to tell you the time on the classroom clock. This relates the skill of telling time to the actual learnt experience of time passing.

More digital times

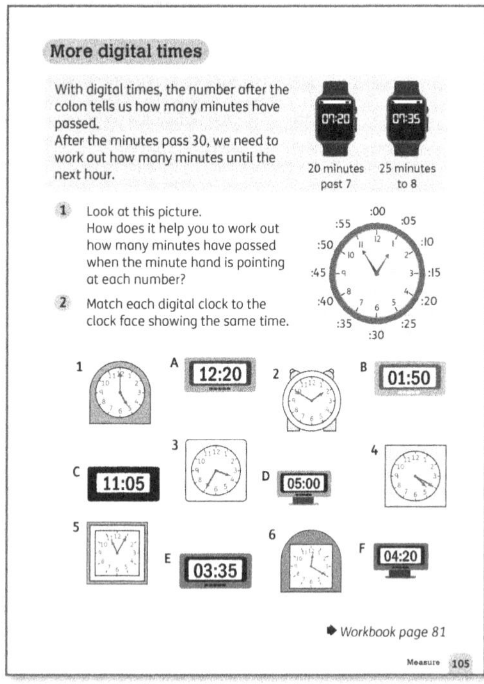

Materials

Digital clock; model digital clock with number cards that can be used to make different times; analogue clock with moveable hands

Warm-up

- Using an analogue clock with moveable hands, show some times on the hour and half-past the hour: 5 o'clock, half past five, and so on. Ask the children to show each time on a model of a digital clock.
- Once the children have represented times on the hour and half-past the hour, move on to times that show quarter past and quarter to.

Focus

- Use the number cards to change the minute numbers on the digital clock model, so you can demonstrate the time for 5-minute intervals (5 past, 10 past, and so on). Count in fives with the class, changing the numbers.
- Read through the information about digital times on **Pupil Book 2 page 105**.

- Ask the children to look at the teaching clock picture in question 1 and discuss what the numbers around the clock face show (minutes past the hour).
- The children can work in pairs to complete question 2.
- Give more examples on the analogue clock face for the children to show on digital clock faces. Keep working practically in this way until they get the idea of showing digital times at 5-minute intervals.

Follow-up

The children complete **Workbook 2 page 81** independently to consolidate their knowledge of how to write digital times.

Challenge

You can give word problems using time intervals to the nearest 5 minutes before or after a given time. For example, ask:

What is the time …
- *half an hour after 8:10 (8:40)*
- *a quarter of an hour after 12:20 (12:35)*
- *10 minutes before 11:40? (11:30)*

Support

It is very helpful to relate the work on time to other maths skills the children have built up over the year. They need to make the connection between:

- skip counting in fives, and the connection between 12 skips of 5 and the 12 numbers around the clock
- counting in ones to 60, and skip counting in fives to 60
- multiplication and division – dividing an hour into chunks of 5 minutes or 15 minutes, and observing how these chunks can be added back together to form the whole
- halves and quarters (learnt in Unit 13), and the concept of a half hour and a *quarter hour*
- the concept of half turns and full turns (which have been introduced at kindergarten level, but will also be covered in the next unit).

The more connections the children can make between these concepts, the more fully they will understand the concepts of telling the time.

Interesting mistakes

You will need to emphasise that once an hour has passed, the time switches to the next hour, so after 5:59 we change to 6:00. The minutes are never shown as 60.

End-of-unit check

Ask questions and carry out practical activities to assess the children's understanding of time, for example:

- Ask the children to show given times on a clock with moveable hands. Do the same with the digital clock.
- Show a time on the digital clock and ask the children to move the hands on their analogue clocks to show the equivalent time, and vice versa (analogue to digital).

Ask questions such as:

- *What do we call it when the minute hand points to the 3?* (quarter past or 15 minutes past)
- *How many quarter turns does the minute hand make around the clock in an hour?* (4)
- *Which is longer: 3 seconds or 3 minutes?* (3 minutes)
- *Can you write your name 100 times in a second?* (No) *Can you do it in a minute?* (no) *Why not? Roughly how long does it take you to write your name?* (Answers will vary depending on how long it takes a child to write his/her name.)
- *How many hours are there in a day?* (24)
- *What time is it on a clock when the big hand points to* (name a number between 1 and 12) *and the little hand points to* (name a number between 1 and 12)?
- *What time is it when the clock reads 10:30?* (half past ten) *What time is it when the clock reads 4:05?* (five past four) *What about 2:35?* (25 minutes to 3)
- *When the time is 04:30, is it half past four or half to four?* (half past four)

UNIT 19 Position and movement

Learning objectives

- Use mathematical vocabulary to describe position, direction and movement, including movement in a straight line and distinguishing between rotation as a turn and in terms of right angles for quarter, half and three-quarter turns (clockwise and anti-clockwise).

Key words

right   left   up   down   straight   direction   turn

quarter turn   half turn   three-quarter turn

full turn   right angle   forwards   backwards

straight   clockwise   anti-clockwise

Unit introduction

Materials

Arrows (cut out of card or paper); masking tape or chalk; toy cars

Teaching guidance

- Play some games in which the children move around the classroom. Give them verbal instructions to get to a location. If any children in your class have limited mobility, you may need to adapt games to make them suitable for everyone.
- To revise the use of position words (prepositions), ask the children to describe positions of objects using words such as *on, in, behind, under, next to*.
- Set out a trail of arrows in a large space (outside, or in the school hall). Include arrows pointing forwards, backwards, left and right. Seat the children at the start of the trail. Choose a child to follow the trail and ask them to say what they are doing as they move along the trail ('I am going forwards', 'I am turning left', and so on). The other children should check that the child describing the trail is correct. Let the children change the route and repeat the process. If they find it challenging to imagine walking the trail, allow them to physically walk it to check their directions. Mark out a simple path including right-angled turns on the floor. The children take turns to walk the path while a partner describes the turns being made, left or right.
- Set up a street plan on the floor using masking tape or chalk. The children take turns giving directions to their partner to follow a route with a toy car.
- The children could also describe the route from the classroom to different places in the school.

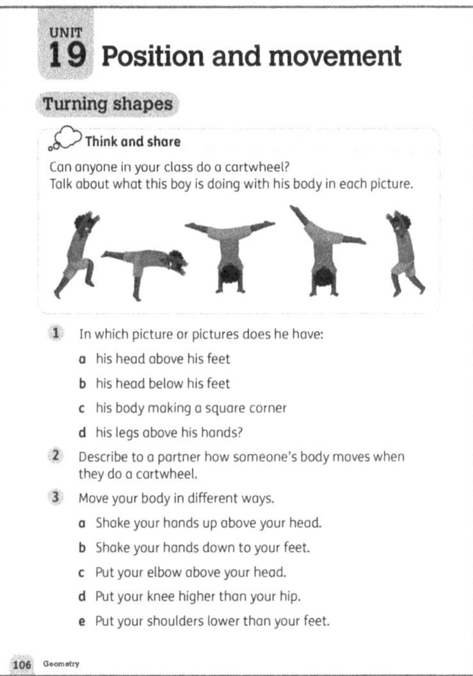

Materials

Objects for demonstrating turns, such as a book, a ruler and cut-out shapes

Warm-up

- <u>Think and share</u>: Discuss the pictures of a boy doing a cartwheel on **Pupil Book 2 page 106** with the class. If you have any skilled gymnasts in the class, they may like to demonstrate cartwheels or tumbles. (If so, take the children to an appropriate area for these activities, where there are suitable mats on the floor.)
- Alternatively, just discuss the pictures. Talk about how the boy in the picture is moving at each stage. Ask:
  - *What is his starting position?*
  - *Which position is halfway between the beginning and the end?*
  - *Which way did he turn?*
  - *If he went the other way, how would he move?*

Focus

- Hold up a book and ask: *How must I move this book to make it do a turn just like the cartwheel?* Begin turning it slightly. After each quarter turn, ask: *Is it back in the starting position? Which way shall I move it now?*
- Ask the children to tell you how they know when the object has made a *full turn* and when it has made a *half turn*.
- Repeat with some other objects, demonstrating turns *forwards* and *backwards*, *left* and *right*, and half turns and full turns. Invite the children to demonstrate with objects and shapes.
- Discuss question 1 with the class.
- Before you discuss question 2, remind the children of the words *left* and *right*. If appropriate for the children in your class, show how the left hand forms an L shape. Another way of teaching left and right is by asking the children to notice which hand they write with. Whether it is left or right, they can usually remember the side of their writing hand and use that to help them remember left and right.
- Work through the practical work in question 3 parts a to c with the class and, if space allows and you have mats, work through parts d and e with the children.

Support

Relate this work on turning to Unit 18 on time. Ask:
- *Which way do the hands of the clock move?* (clockwise/to the right)
- *How long does it take the minute hand to make a full turn around the face of the clock?* (60 minutes/1 hour)
- *How long does it take to make a half turn?* (30 minutes) *How long does it take to make a quarter turn?* (15 minutes)
- *At what time has it made a full turn?*

Answers for Pupil Book 2 page 106

<u>Think and share:</u> Individual's answers. For example: the boy raises his arms and one leg; he makes a rotation with one of his feet in the air, he puts both hands in the floor and his legs in the air and then moves back to the starting position.

1 a first and fifth pictures
   b third and fourth pictures
   c second and fourth pictures
   d third and fourth pictures
2 Individual's answer
3 Practical work

Moving in a straight line

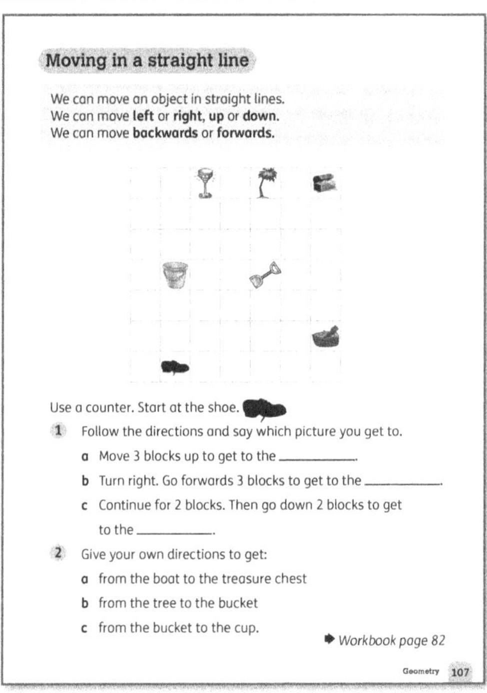

Materials

Counters

Warm-up

Start with this open-ended puzzle.

- Draw four dots on the board, arranged like the corners of a square.

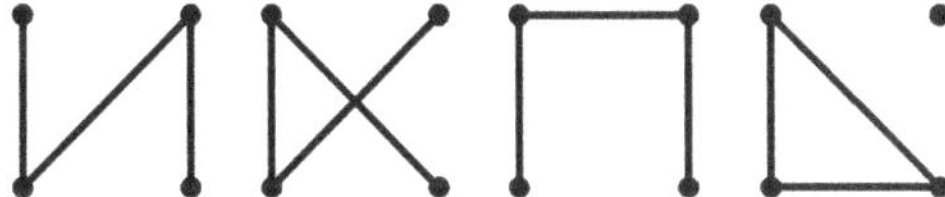

- Ask: *What shape can I make if I join the dots only moving my chalk in straight lines?* The most obvious answer will be a square. However, extend the activity to say: *What other shapes can you make? The only rules are that you must move the chalk in straight lines and not go over any lines more than once. But you can make as many lines as you like and the shape does not have to be closed.*
- Here are a few different solutions:

- Invite some children to come up and try to make a variety of different shapes. You could try the same exercise with three rows of three dots.

Focus

Encourage use of the vocabulary *left, right, up, down, straight, forwards* and *backwards*. Discuss these questions with the class:

- *What do you think a straight line is?*
- *How is a straight line different from a different kind of line?*
- *What does it mean when you move in a straight line?*
- (while walking sideways) *I am moving in a straight line. In which direction am I moving?*
- (changing to the opposite direction) *Now in which direction am I moving?*
- *Who can give me directions to get to another place in the classroom only moving in straight lines?*

If you wish, you can play a game in an open space such as the school hall or outside:

- Give the children directions to move 1, 2 or 3 steps forwards, backwards, right or left.
- If any children in your class have limited mobility, you may want to adapt this game. Give the children directions to move objects on a tabletop.

Finally, discuss the information on **Pupil Book 2 page 107**. Then work through question 1 and question 2 with the class.

Follow-up

Each grid on **Workbook 2 page 82** shows one or two counters and an instruction. The child follows the instruction to move the counter (or counters) to the new position. Demonstrate with the first one and then let the children complete the page independently.

Challenge

A fun game that the children can play is 'Battleships'. This is excellent practice for movement and direction work. You can find instructions for this online.

Interesting mistakes

Children sometimes find the use of the words up and forwards confusing. On a flat map, an upward-pointing arrow indicates north. Up can also mean forwards. However, up and down mean something different in real life, where we talk about moving higher and moving lower. Discuss these differences with the class so that you can reach a consensus about the best way to use direction words when dealing with maps.

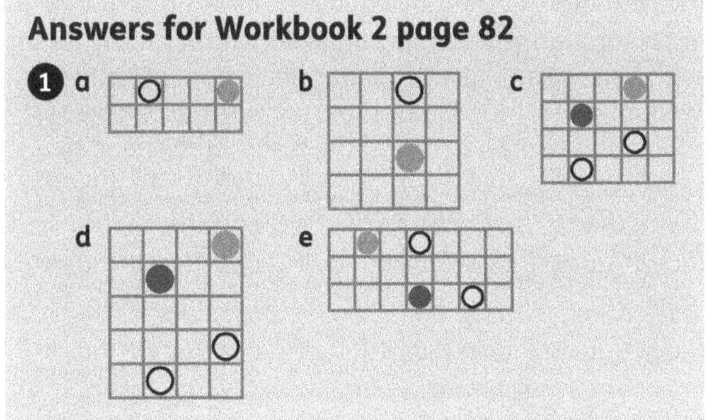

Answers for Pupil Book 2 page 107

1   **a**   bucket   **b**   spade   **c**   boat

2   Possible answers:
- **a**   Move 5 blocks up.
- **b**   Move 3 blocks down. Turn right. Go forwards 3 blocks.
- **c**   Move up 3 blocks. Go forwards 1 block.

Answers for Workbook 2 page 82

1   **a**   **b**   **c**   **d**   **e**

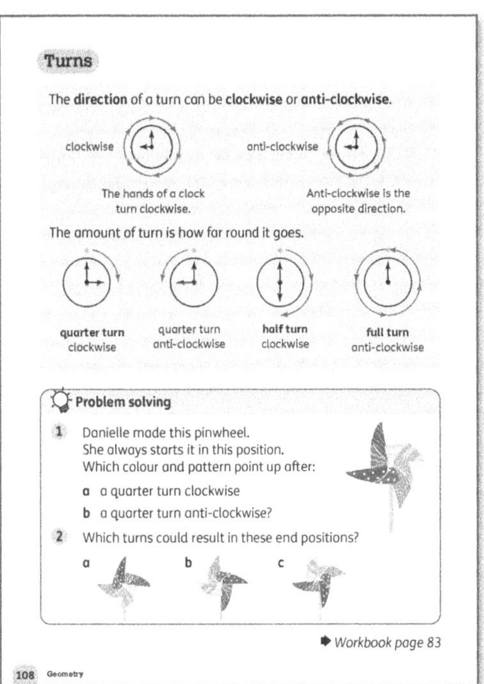

Turns

Materials

Analogue clock faces with moveable hands; jars or bottles with screw lids; paper plate with an arrow to show turns (optional)

Warm-up

- Revise the terms *direction*, *clockwise* and *anti-clockwise* with the class. Demonstrate clockwise and anti-clockwise movements on the clock faces.
- If each child still has a moveable clock face, ask them to move the hands in the correct direction as you call out clockwise or anti-clockwise. Alternatively, they can demonstrate the direction by moving their hands or arms in that direction.
- Demonstrate how we open or close the lid of a jar. Ask the children to identify which way you turn the lid to open it and which way to close it.

Focus

- Work through the information on **Pupil Book 2 page 108** pointing out the meaning of *quarter turn*.
- Problem solving: Work through question 1 and question 2 with the class.

Follow-up

Work through **Workbook 2 page 83** with the class (or with smaller groups). Use a clock face or paper plate with an arrow pointing to show the turns as necessary. Allow the children to model the turns if they need to.

Answers for Pupil Book 2 page 108

Problem solving:

1   **a**   red with white dots   **b** yellow stripes

2   **a** $\frac{1}{4}$ turn clockwise

    b $\frac{1}{2}$ turn clockwise or anti-clockwise

    c $\frac{1}{4}$ turn anti-clockwise

Answers for Workbook 2 page 83

1   a 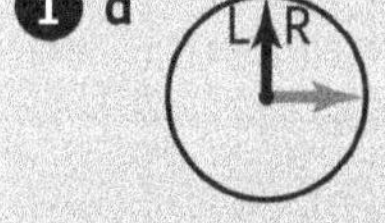   b 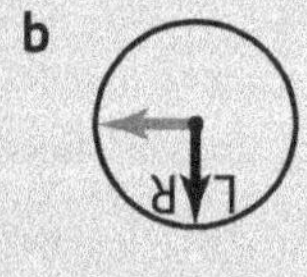   c 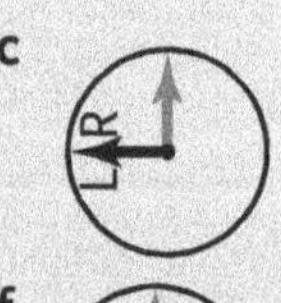

    d 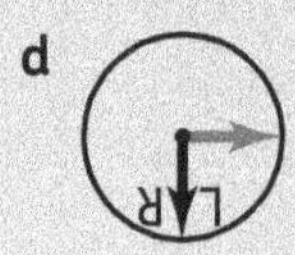   e 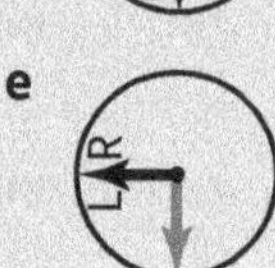   f 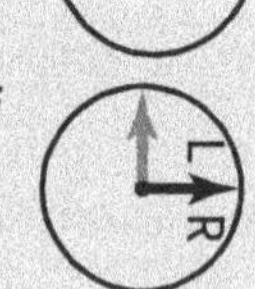

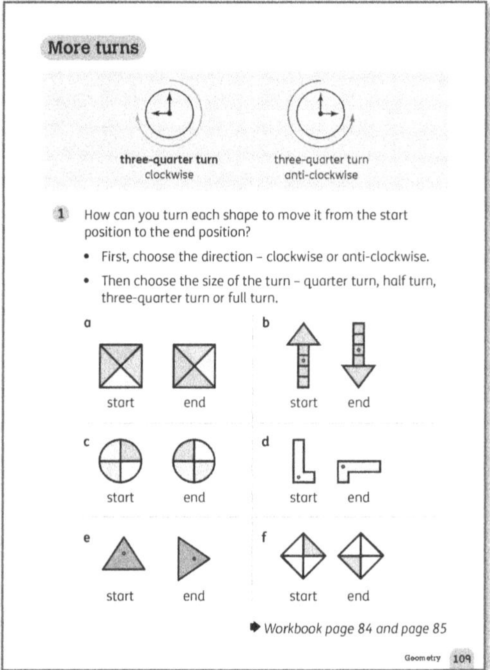

Materials

Angles made from cardboard strips and paper fasteners; squared paper for demonstrating shape turns; a simple shape with line symmetry to rotate on the grid

In this lesson, the children focus on rotating shapes. They need to focus on:
- the size of the turn – one-quarter, one-half, three-quarters and a full turn
- the direction of the turn – clockwise and anti-clockwise.

Set up an 'angles area' of the classroom with various materials that the children can use to explore angles.

Warm-up

- Give the children two strips of paper joined by a split pin. Each strip forms an 'arm' that can open to increase the size of the angle at the join. Ask the children to use the paper angles to demonstrate the movement of the two strips through a quarter turn, half turn, *three-quarter turn* and full turn.

Focus

- Fold a large piece of paper to make a right angle and remind the children of the term *right angle*. Relate the turns in the grid from the previous lesson to your right angle.

The children do not formally have to know the term *right angle* yet – you can refer to square corners. However, if you wish, you can ask: *Does anyone know what we call this kind of corner?*

- Ask the children to find examples of right angles in the classroom and encourage them to name shapes that contain right angles.
- Demonstrate turning a straight object (a pen or ruler) through quarter turns to make a full turn.
- Once the children have a sense of what each size of turn means, place a cut-out shape with a single line of symmetry onto a piece of squared paper. Demonstrate moving the shape through 4 quarter turns and point out how it changes position each time before returning to its original position. You can use the pictures on **Pupil Book 2 page 109** for suggestions of shapes and designs that can help the children to notice the way a shape turns through each position as it rotates.
- Discuss the information on the Pupil Book page before working through question 1 with the class.

Follow-up

Let the children complete question 1 on **Workbook 2 page 84** independently and then have them compare their answers with a partner. Let them discuss their solutions if they get different numbers of turns and encourage them to reach agreement.

They can work on questions 1–3 on **Workbook 2 page 85** in pairs, then compare answers with another pair and correct each other's answers as needed.

Interesting mistakes

Strictly speaking, a mathematical rotation refers only to the turning of a shape around a fixed point. In order to illustrate the turns on a page, the start and end positions of each shape are shown. However, some children will notice that it appears as though the shape has slid to a new position on the grid. For this reason, it is important to demonstrate the rotations using real cut-out shapes that turn about a fixed point.

Answers for Pupil Book 2 page 109

1. a $\frac{3}{4}$ turn clockwise or $\frac{1}{4}$ turn anti-clockwise
   b $\frac{1}{2}$ turn clockwise or anti-clockwise
   c $\frac{3}{4}$ turn clockwise or $\frac{1}{4}$ turn anti-clockwise
   d $\frac{1}{4}$ turn clockwise or $\frac{3}{4}$ turn anti-clockwise
   e $\frac{1}{4}$ turn clockwise or $\frac{3}{4}$ turn anti-clockwise
   f $\frac{1}{2}$ turn clockwise or anti-clockwise

Answers for Workbook 2 page 84

1. hen: 5 quarter turns; rabbit: 5 quarter turns
   monkey: 7 quarter turns; cat: 3 quarter turns

Answers for Workbook 2 page 85

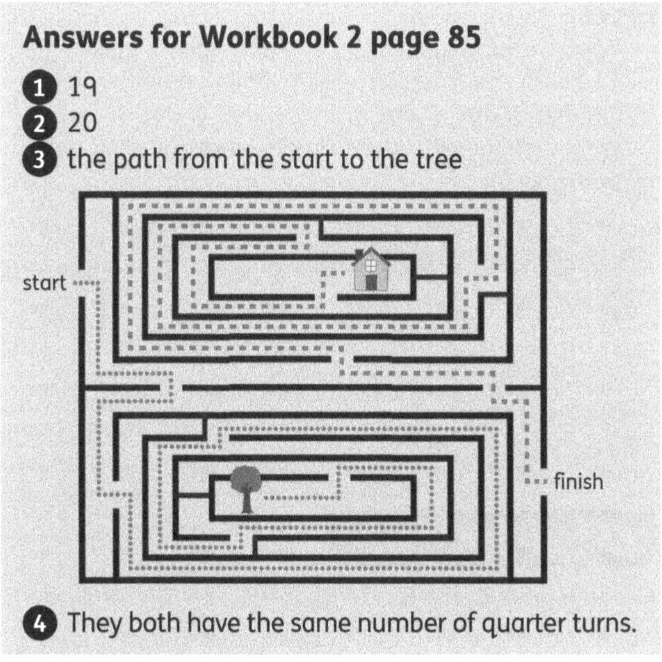

1. 19
2. 20
3. the path from the start to the tree
4. They both have the same number of quarter turns.

Working with turns

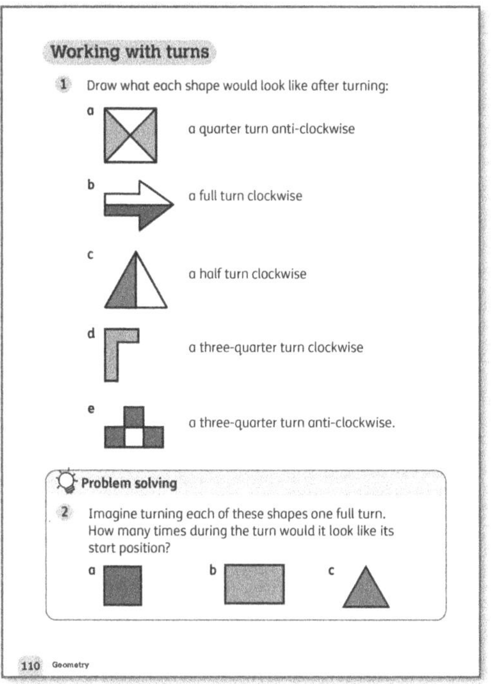

Materials

A shape with a single line of symmetry

Warm up

Demonstrate turning a shape with a single line of symmetry through turns to make a full turn, first going clockwise and then going anti-clockwise. Then call out a turn, for example, a quarter turn anti-clockwise and encourage the children to predict what the shape will look like with such a turn. Repeat for different turns, encouraging the children to make predictions each time for what the shape will look like when turned.

Focus

- Work through question 1 on **Pupil Book 2 page 110** with the class, letting the children work independently if they are confident with the task.
- <u>Problem solving</u>: Let the children work in pairs on question 2.

Answers for Pupil Book 2 page 110

1 a   b   c   d   e

<u>Problem solving:</u>
2 a 4 times **b** 2 times **c** 3 times

End-of-unit check

Ask questions to assess the children's understanding of turns. For example:

- *How many quarter turns make a full turn?* (4)
- *How many half turns make a full turn?* (2)

- (While showing a shape with symmetry) *How far must I turn this shape before it returns to its starting position?* (Answer depends on the shape.)
- *Can you turn this shape a quarter turn clockwise/ a quarter turn anti-clockwise/a full turn clockwise?* (The child demonstrates turning the shape a quarter turn clockwise/a quarter turn anti-clockwise/a full turn clockwise.)
- *What does clockwise mean?* (Turning in the same direction that the hands turn). *Can you show me which direction is clockwise?* (For example, child uses their arms to demonstrate a clockwise turn or turns a shape clockwise.)

You could give the children verbal directions to draw a shape or letters that involve right angles, for example:

- *Draw a dot. Draw a line 4 cm going up from your dot. Then make a quarter turn to the right. Draw a line that goes 2 cm to the right. Lift up your pencil and go back to the first dot. Move 2 cm up from the dot, make a quarter turn right and draw a 2 cm line out to the right. What letter have you made?* (a capital F)
- Repeat with other numbers or letters.

UNIT 20 Money

Learning objectives

- Recognise value and money notation used in local currency.
- Recognise and use symbols for pounds (£) and pence (p).
- Combine amounts to make a particular value.
- Find different combinations of coins that equal the same amounts of money.
- Solve simple problems in a practical context involving addition and subtraction of money of the same unit, including giving change.

Key words

coins notes money currency cash value combination worth change

Unit introduction

Teaching guidance

Encourage the children to talk about their experiences with *money*. This allows you to introduce or reinforce the terms *coins* and *notes* and to see how much they already know. Ask the children to say what they think is a large amount of money and what they think is a small amount of money.

A generic form of cents and dollars is used in this unit. You will need to support this work using the coins and notes from your own country. If you do not use cents and dollars, explain to the children that many countries use cents and dollars, and that it is helpful to be able to work with these currencies.

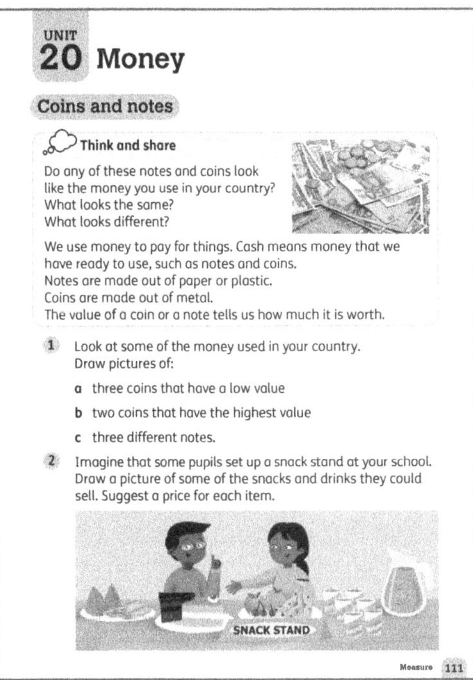

Materials
Example coins and notes from your country (real or mock-ups; enough real coins for at least one per child)

Warm-up
- Pass the coins around and ask the children to describe the coins as they handle them. For example: *This is a one-cent coin, it is brown and it has a picture of a hummingbird on it.* Repeat this for all the coins, giving different children a chance to describe them. Use the words coins and cents regularly (or local currency unit instead of cents). Make sure the children can read the word 'cents' (or local equivalent) on the coins and can find the name of their country on the back of the coin (if applicable). Also use the term *cash*.
- Talk about the symbols used on the coins and discuss how they help us to recognise our own money.
- Children can make some coin rubbings: they place a piece of paper over a coin and shade it lightly with a pencil. The children could cut these out to make pretend coins to use in class.
- Introduce the notes. Discuss how the colour and symbols help us to recognise the different amounts and stress that different notes have different values. Let the children make rough sketches of some notes.

Focus
Keep in mind that the addition and subtraction requirements at this stage only involve pairs of 1- and 2-digit numbers, although the children can add groups of small numbers. Keep to whole numbers for calculations and not decimal values.

- Put the children in groups and give each child a coin. Ask them to discuss their coins in groups. They should tell each other what their coin is called, how much it is *worth* (its *value*), what they could buy with it, what they could not buy with it, and so on.
- Describe a coin to the class for the children to guess which coin you are describing. Say things such as: *It is about this big. It is brown. It has a picture of a hummingbird on it. What is it?*
- Think and share: Discuss the picture of coins and notes on **Pupil Book 2 page 111**.
- For question 1, hand out some coins and notes from your country (either real or mock-ups). Let the children discuss the values of the coins and notes.
- Discuss question 2 with the class and let them share their ideas.

Interesting mistakes
- The children may confuse the concept of dollar bills or notes with dollar coins, depending on what is in use locally. Explain that some countries use coins while others use notes. Some children may regard coins as more valuable, as metal is heavier and may appear or feel more precious than paper.
- Some children may have difficulty remembering which coin is which. They may also find it difficult to grasp the concept of the value of a coin or note. Help them by talking about which can buy more.

Answers for Pupil Book 2 page 111
Think and share: Individual's answer
1 Individual's answer
2 Individual's answer

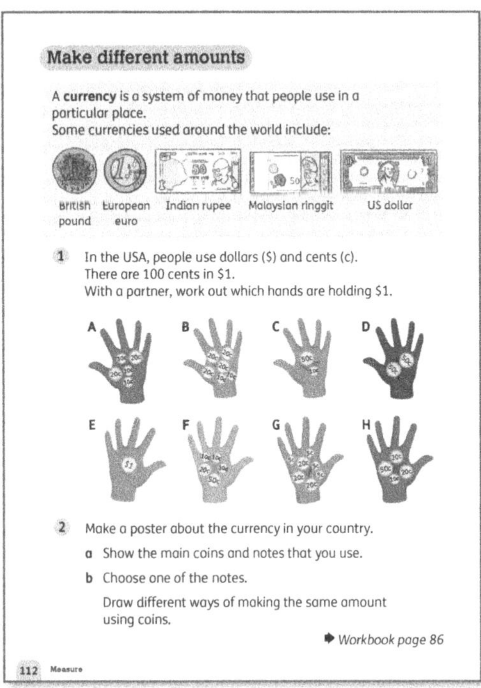

Materials

Coins (real or pretend), such as 5c and 1c coins (or adapt as suitable for your region); several stones wrapped up in three colours of paper (or buttons in different colours); large sheets of paper and markers

Warm-up

- Give each group a pile of 5c and 1c coins. Tell them that each group member is to be given 10 cents. Let them make this in any way they can and discuss the different ways (ten 1c; five 1c + one 5c; two 5c; and so on). Introduce the term *combinations*. Ask how else we could get 10 cents (one 10c).
- Use the coloured stones to make a shop. Explain that the stones are different prices (red is 1c, blue is 5c and green is 10c, for example). Give each child 15c to spend. Let them work out all the possible combinations of stones they could buy. Next, tell them a blue stone costs 15c. Let them draw all the ways in which they could pay for it using coins.
- Give each child 5c. Draw a sweet on the board and write 7c next to it. Ask the children how much more money they need to buy the sweet. Repeat this several times with different starting amounts and different prices. Once you are sure the children can work out the difference, repeat this with a lower amount and check that they understand the term *change*: ask how much change they will get. Say, for example: *You have 5c and you want to buy a toffee for 3c. How much change will you get?*
- Give each child a 'purse' with some paper coins in it (1c, 5c and 10c). They take turns to be the shopkeeper. Give the shopkeeper the stones to sell. The other children take turns to buy stones and pay for them. The shopkeeper can give change.

Focus

- Discuss the information and questions on **Pupil Book 2 page 112**, introducing the term *currency*.
- For question 1, discuss the different ways of making up $1.
- For question 2, the children can work in groups or as a class to make a poster about money in your country. If you wish, give each group a particular note or coin and they can draw pictures of other sets of coins or notes that make the equivalent amount. You can use these to build up a classroom display.

Follow-up

- On **Workbook 2 page 86**, the children draw local notes and coins. Use this as an opportunity to discuss how much everyday items cost in your area.

It is likely that the money amounts shown on this page are unrealistic. You can use this page as a starting point for giving the children similar questions using your own currency in more realistic amounts. However, remember that, at this stage, calculations should be kept to amounts under 100 as the children have not yet worked with 3-digit calculations.

Support

Some children may initially not see that combinations of coins can be worth the same as one coin of a different amount. Practical activities in which they build up amounts using smaller denominations will help them to develop this concept.

Answers for Pupil Book 2 page 112

1. B, D, E, F, H
2. Individual's answer

Answers for Workbook 2 page 86

1. Individual's answer
2. Individual's answer
3. Individual's answer

Pounds and pence

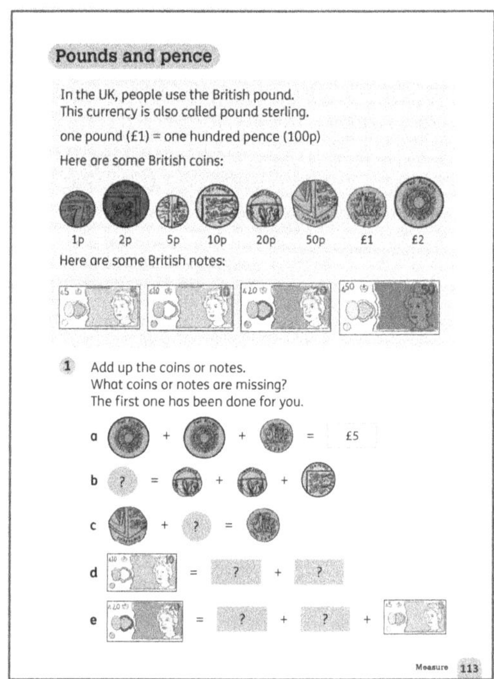

Materials

Mock-ups of coins and notes in pounds and pence

You can find images online or draw and photocopy your own versions of the coins and notes shown here:

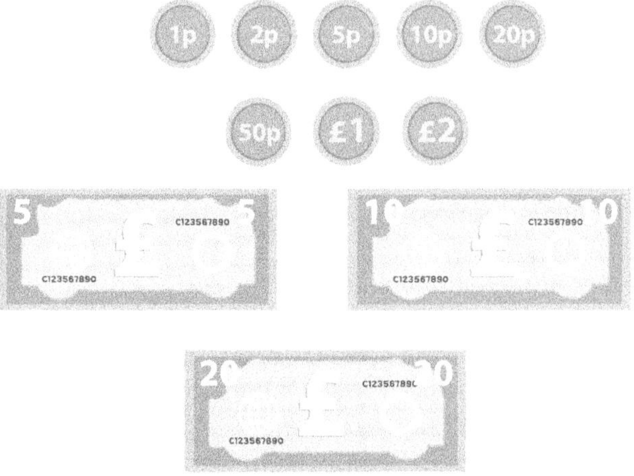

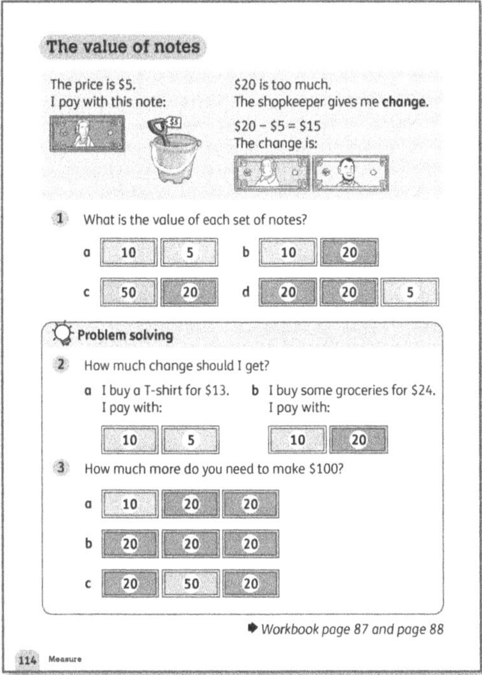

Warm-up

Introduce the terms *pounds* and *pence* and explain that these are the currency units used in the UK. If any children from your class have travelled to the UK or used pounds and pence, invite them to share their experiences. Show them the pound symbol (£) and explain that 100 pence is the same as 1 pound.

Focus

Show the examples of the British coins and notes. Discuss similarities and differences between the British currency and your home currency. Here are some questions and prompts you could use to guide the discussion:
* *How much is this coin worth?*
* *Can you find a coin that is worth more/less than this one?*
* *Which coin has the highest value?*
* *Which is the smallest coin?*
* *Which coin has an unusual shape?*
* *Sort the coins into brown and silver coins.*
* *Sort the coins from smallest to biggest.*
* *Arrange the coins from smallest to greatest value.*
* *Which coins can make the same value as this coin?*
* *Which coins can make the same value as this note?*

Work through question 1 on **Pupil Book 2 page 113** with the class, using questions and prompts similar to the ones given above.

Challenge

Provide some more challenging problems involving pounds and pence. Ask, for example: *What is the biggest/smallest value you can make using 2 coins and 2 notes? How many different ways can you make £20 using £1, £2 and £5 coins? How many ways can you make £20 using notes?*

Answers for Pupil Book 2 page 113

1 **a** £5 [Provided as an example]
   b 50p   **c** 50p
   d £5 + £5   **e** £10 + £5

Materials

Examples or copies of local currency notes; photographs of notes from different currencies

Warm-up

Show and discuss the different notes used in your country. The children can describe the colour and pictures that help us to distinguish them. Here are some questions you could ask to guide the discussion:
* *What do you understand by value?*
* *Which values of notes do we use here?*
* *Which is the highest value?*
* *Which is the smallest value?*
* *What colours/pictures do you see on this note?*
* *Where can we see the value?*
* *What is the value of the amount of these two notes together?* (Repeat with various combinations.)
* *How would I make $20 from two different notes?* (Repeat for $50 and $60, for example.)

Focus

* Turn to **Pupil Book 2 page 110** and work through question 1 with the class .
* <u>Problem solving:</u> Work through question 2 and question 3. Some children may be able to complete the work independently.

Follow-up

* **Workbook 2 page 87** gives the children an opportunity to calculate using amounts of money.
* **Workbook 2 page 88** presents more problem-solving tasks in a money context. Discuss the key with the children and what each pattern represents, then work through question 1 with them. Some children will be able to work independently through the rest; others will require more assistance.

Answers for Pupil Book 2 page 114

1 a $15   b $30
   c $70   d $45

Problem solving:
2 a $2   b $6
3 a $50   b $40   c $10

Answers for Workbook 2 page 87

1 5 cents, 8 cents, 3 cents, 11 cents, 7 cents
2 $9, $3, $5, $8, $16, $40

Answers for Workbook 2 page 88

1 The numbers in each addition can be in any order:
   a $20 + 5 + 1 + 20 = 46c$ b $10 + 5 + 1 + 20 = 36c$
   c $1 + 5 + 5 + 2 = 13c$ d $1 + 20 + 10 + 2 = 33c$
2 Patterns that show these amounts (with the blocks in each pattern in any order):
   a $20 + 10 + 5 + 2 + 1$
   b $20 + 20 + 5 + 1 + 1$ or $20 + 10 + 10 + 5 + 2$

End-of-unit check

Ask questions to assess the children's understanding of money, for example:
- *What coin is this* (Show a coin)*? How do you know?* (For example, the child identifies coin using colour/metal/size/number.)
- *Is this coin from* (your country)*? How do you know?* (For example, the child shows the local currency on the coin or the head of state of that country.)
- *What note (bill) is this?*
- *Is this note (bill) from* (your country)*? How do you know?* (For example, the child shows the local currency on the note (bill) or the head of state of that country.)
- *What is the value of this coin* (Show a coin)*? How do you know?* (For example, the child uses colour/metal/size/number to identify it.)
- *What can I buy with this coin?* (For example, the child identifies items they can purchase locally with the coin, such as sweets.)

- *How many … cent coins make … cents?* (For example, the child identifies the multiple needed to make a specified amount using one type of coin.)
- *What coins can I use to make … cents?* (For example, the child demonstrates how to combine coins to make a specified amount.)
- *I have … I have to pay … How much more money do I need?* (For example, the child demonstrates how much more money and which coins to use to make a specified amount.)
- *I have … I pay … How much change do I get?* (For example, the child demonstrates how much change would be given if they bought something costing a certain amount with a specified amount of money, such as $0.10 change from purchasing something costing 90 cents using $1.)

Mixed practice 3

Answers to Mixed practice 3 on Pupil Book 2 pages 115–116

1 a D; there is no line of symmetry.
   b Possible answers (each of these shapes has more than one line of symmetry):

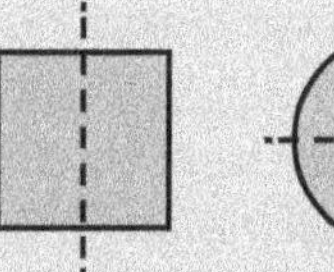 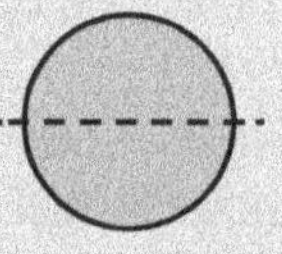 

2 box of milk: 1 litre; cup: less than 1 litre
   spoon: less than 1 litre; bottle of oil: 1 litre
   water bottle: more than 1 litre
3 a Individual's answers
   b an instrument that we use to measure temperature
4 a 20 (minutes) past 2   b 12 o'clock
   c 5 (minutes) to 11   d 10 (minutes) to 2
5 a down   b right   c up
6 a $20   b $30   c $95
   d $85   e $70   f $100
7 a $30   b $30   c $15